Nuclear, Chemical, and Biological Terrorism

Emergency Response and Public Protection

Mark E. Byrnes
David A. King
Philip M. Tierno, Jr.

LEWIS PUBLISHERS

A CRC Press Company
Boca Raton London New York Washington, D.C.

Library of Congress Cataloging-in-Publication Data

Catalog record is available from the Library of Congress

Lewis Publishers is an imprint of CRC Press LLC

International Standard Book Number 1-56670-651-3
Printed in the United States of America 1 2 3 4 5 6 7 8 9 0
Printed on acid-free paper

Foreword

The authors of this excellent book provide a concise but comprehensive review of various types of weapons of mass destruction, along with sound advice and simple actions that can be taken by emergency responders and the general public to reduce risks and avoid panic in the event of a terrorist attack. By simply reading through this book, emergency responders and the public will learn what they can do to minimize danger to health and life after an attack. The most important actions are summarized at the end of the book. This summary becomes a convenient checklist.

Through my own personal experiences facing high intensities of radiation and radioactive materials, managing patients who have been exposed to intakes of radioactive material, and training and discussing concepts and actions with emergency responders and the public, I've found that people want to do the right thing when faced with dangerous situations. In the event of a terrorist attack, they want to respond by protecting health and saving life. This book helps them do that.

I recommend that emergency responders and persons establishing homeland security programs read this book, along with every person who wants to conserve health and save life. The more people that know the material in this book, the less will be the panic and loss of life in the terror that follows a weapon of mass destruction attack. There are other books on this subject that are less complete and largely out-of-date. This book is comprehensive, up-to-date, and provides sound advice for protection in the event of a terrorist attack involving weapons of mass destruction.

Allen Brodsky, Sc.D., CHP, CIH, DABR
Adjunct Professor of Radiation Science
Georgetown University Chair
Ad Hoc Committee on Homeland Security
Health Physics Society

Preface

Following the events of September 11, 2001, the United States began a strategic assault against terrorist groups around the world with the objective of ridding the world of large terrorist organizations that could potentially repeat a September 11 type event. As an added level of security, President George W. Bush established a new Department of Homeland Security that has the responsibility for reducing America's vulnerability to terrorism. While both actions represent good first steps in helping rid the world of terrorism, the American public has been forced to come to grips with the reality that the nation faces a strong possibility of future large-scale terrorist events occurring within its borders. The American public has also come to realize that it must play a critical role in helping prevent these types of events.

While everyone can gain valuable information from this book on how to protect themselves from terrorist activities involving nuclear, chemical, and biological weapons, its target audiences are emergency response personnel, safety professionals, law enforcement officials, and Federal Bureau of Investigation agents (all referred to as emergency responders) because they are likely to receive first-hand exposure to one or more terrorist events involving these types of weapons. The primary objectives of this book are to provide emergency responders with guidance on:

- Weapons of mass destruction that could be used in a terrorist attack
- Mechanisms by which terrorists could disperse various types of nuclear, chemical, and biological agents
- Conventional explosives that terrorists could use to disperse these agents
- Routes by which individuals are exposed to these agents
- Health hazards that may result from exposure to these agents
- Techniques by which safety professionals can minimize exposure to these agents
- Potential medical treatment options for those exposed to these agents
- Methods to increase chances of surviving a nuclear explosion
- Emergency preparedness measures for a variety of settings
- Techniques for prioritizing injuries
- Personnel decontamination methods to be administered prior to medical treatment
- Radiation exposure guidelines
- Training guidelines

It is the authors' intent that the information in this book will help reduce exposure of emergency responders to these types of agents, and as a result, help save lives.

Authors

Mark E. Byrnes, P.G. is a senior scientist at the U.S. Department of Energy's Hanford Nuclear Reservation in Richland, Washington. Byrnes works for Fluor Hanford and has 18 years of experience performing and designing environmental investigations in radioactive and chemical (including warfare agent) environments. He earned a B.A. in geology from the University of Colorado at Boulder and an M.S. in geology and geochemistry from Portland State University in Oregon. Byrnes is an adjunct professor at Washington State University's Tri-Cities Campus and is a registered professional geologist in the states of Washington, Tennessee, and Kentucky. He is the author of two textbooks used at major universities: *Sampling and Surveying Radiological Environments* and *Field Sampling Methods for Remedial Investigations*, published by CRC Press and Lewis Publishers in 2001 and 1994, respectively.

David A. King, C.H.P. is a certified health physicist working for Science Applications International Corporation in Oak Ridge, Tennessee. King earned a B.S. in physics from Middle Tennessee State University (1991) and an M.S. in radiation protection engineering from the University of Tennessee at Knoxville (1993). He received certification from the American Board of Health Physics in 1999 (recertified in 2003) and is a member of the Health Physics Society and the Society for Risk Analysis. King's primary responsibilities include managing environmental studies for radiologically contaminated sites, designing characterization plans and interpreting associated data, preparing CERCLA documentation, and conducting human health risk and dose assessments. Major clients include the U.S. Department of Energy on the Oak Ridge Reservation (Oak Ridge, Tennessee) and in Paducah, Kentucky and the U.S. Army Corps of Engineers through the Formerly Utilized Sites Remedial Action Program (FUSRAP).

Philip M. Tierno, Jr., Ph.D. is a well known microbiologist with more than 30 years of experience in the fields of clinical and medical microbiology. He is the director of clinical microbiology and diagnostic immunology at Tisch Hospital, New York University Medical Center, as well as Mt. Sinai Medical Center, and is a part-time associate professor at the New York University School of Medicine. He performed his graduate studies at New York University, where he was awarded an M.S. in 1974 and a Ph.D. in 1977. Dr. Tierno acts as a consultant to the office of the attorney general of New York State, the Department of Health of the City of New York, the National Institutes of Health in Bethesda, Maryland, and the College of American Pathologists. Dr. Tierno is a member of the New York City Mayor's Task Force on Bioterrorism. Technical articles written by Dr. Tierno have appeared in the *American Journal of Public Health*, *Journal of Clinical Microbiology*, *American Journal of Clinical Pathology*, *Reviews of Infectious Diseases*, *Journal of Infectious Diseases*, and other publications. In the past 10 years, he has also authored several books including *Staying Healthy in a Risky Environment* (Simon & Shuster), *The Secret Life of Germs: Observations and Lessons from a Microbe Hunter* (Pocket Books), and *Protect Yourself against Bioterriorism* (Pocket Books).

Acknowledgments

The authors would like to recognize Richard Wilde, P.K. Brockman, and Donald Moak (Duratek Federal Services) for providing funding to support the editing of this book as well as technical resources with expertise in the areas of explosives and explosives transportation.

We would also like to recognize Kenny Flemingh (C.H.P.), Allen Brodsky (C.H.P.), and Theresa Patterson (P.M.P.) from Science Applications International Corporation; Richard Toohey (C.H.P.) of the Oak Ridge Institute for Science and Education; Phillip Amundson (C.I.H., C.S.P.), Manager of Safety and Emergency Preparedness for Tacoma School District 10; Bruce Cannard (Principal, Canyon View Elementary School); and Rose Ruther for performing technical reviews on multiple sections of this book and providing technical expertise in the areas of radiation health physics, radiation safety, safety engineering, and industrial hygiene. Microecologies, Inc. should also be recognized for financial and technical contributions in the area of industrial hygiene.

The authors would like to express their appreciation for all the support provided by their families, particularly Karen Byrnes, Christine Byrnes, Kathleen Byrnes, Frieda Byrnes, and Shelley King, Josephine Tierno, Alexandra and Francois Payard, and Meredith and Thomas Mallon.

Contributors

Grant M. Ceffalo, C.H.P. is the manager of the Radiological Control Department for Bechtel Hanford, Inc. at Richland, Washington. Ceffalo has worked in the fields of radiation monitoring instrumentation, dosimetry, environmental measurement and release, as well as training. He earned a B.S. in radiation science from the University of Washington and an M.S. in radiological science from Colorado State University. Ceffalo is a comprehensively certified health physicist and a member of the American Academy of Health Physics.

Sheldon R. Coleman, C.I.H., P.E. is the industrial hygiene program administrator for Bechtel Hanford, Inc. at the Department of Energy's Hanford site at Richland, Washington. He has extensive experience in respiratory protection, chemical hazard evaluation, and the design of engineering controls. He also has 10 years of experience with the U.S. Air Force in the fields of nuclear, biological, and chemical warfare. Coleman earned a B.S. in chemical engineering from the University of Washington, attended the U.S. Air Force School of Aerospace Medicine, and earned a master's degree in public administration from the University of Oklahoma. He is certified in the comprehensive practice of industrial hygiene and registered as a professional engineer.

William M. Sothern is an industrial hygienist and chief investigator for Microecologies, Inc., a New York City-based firm that conducts environmental investigations and performs environmental cleanup contracting. Sothern holds a master's degree in environmental and occupational health from Hunter College in New York and is a member of the American Industrial Hygiene Association and the American Conference of Government Industrial Hygienists. He has been involved closely with the air monitoring and cleanup efforts in the downtown New York area in the aftermath of the World Trade Center disaster and is currently working with New York City companies and residents to develop emergency preparedness plans for possible future terrorist attacks.

Richard P. Genoni is a principal engineer with Duratek Federal Services' Northwest Operations in Richland, Washington. Genoni maintains the Explosive Classification Tracking System for the Department of Energy's National Transportation Program and reviews all new explosive applications before submittal to the Department of Transportation. He also maintains Department of Energy Interim Hazard Classifications in accordance with the Department of Defense Ammunition and Hazard Classification Procedures, TB 700–2.

Diane Forsyth is a technical editor at Duratek Federal Services' Northwest Operations in Richland, Washington. In addition to performing the primary editing of this book, she has authored and edited numerous articles, newsletters, books, and other printed resources. She earned a B.A. from Walla Walla College, Washington, and an M.A. from Andrews University, Berrian Springs, Michigan. She has also done extensive doctoral work at schools in California and Washington, D.C.

Scott D. Elliott is manager of the Waste and Transportation Services Department for Duratek Federal Services at the Hanford Nuclear Reservation, Richland, Washington. Elliott has worked in the fields of nuclear operations, training, and waste treatment for 24 years and has additional experience as a nuclear facility building emergency director. He is presently a nuclear, biological, and chemical operations instructor for the U.S. Army Reserves.

Contents

Dedication

This book is dedicated to the emergency responders who lost their lives in the September 11, 2001 terrorist attack, as well as their families.

It is also dedicated to my father, Francis J. Byrnes, who taught me to enjoy and appreciate the fields of science and engineering, and provided me with guidance and encouragement throughout my professional career.

MEB

Abbreviations and Acronyms

AC Hydrogen cyanide (blood agent)
ALARA As low as reasonably achievable
BAL British Anti-Lewisite
CIA Central Intelligence Organization
CK Cyanogen chloride (blood agent)
CX Phosgene oxime
DFSH Duratek Federal Services Hanford
DFSNW Duratek Federal Services Northwest
DNA Deoxyribonucleic acid
DNT Dinitotoluene
ECD Electron capture detector
ED Ethyldichloroarsine (arsenical blister agent)
EGDN Ethylene glycol dinitrate
FH Fluor Hanford
FID Flame ionization detector
g Gram
GA Tabun (nerve agent)
GB Sarin (nerve agent)
GC Gas chromatography
GD Soman (nerve agent)
HAZMAT Hazardous materials
HD Distilled mustard
HEPA High efficiency particulate arrestor
HL Mustard–lewisite mixture
HMX Cyclotetramethylenetetranitramine
HN-1 2,2-Dichlorotriethylamine (nitrogen mustard agent 1)
HN-2 2,2-Dichloro-N-methyldiethylamine (nitrogen mustard agent 2)
HN-3 2,2,2-Trichlorotriethylamine (nitrogen mustard agent 3)
HNS Hexanitrostilbene
HPLC High-performance liquid chromatography
HT Distilled mustard combined with bis(2-chloroethyl sulfide) monoxide
IDLH Immediately dangerous to life or health
kg Kilogram
L Lewisite (arsenical blister agent)
lb Pound
LMNR Lead mononitoresorcinate
LNT Linear no threshold
KDNBF Potassium dinitrobenzofurozan
KGB Komitet Gosudarstvennio Bezopaznosti

MD Methyldichloroarsine (arsenical blister agent)
NATO North Atlantic Treaty Organization
PD Phenyldichloroarsine (arsenical blister agent)
PETN Pentaerythritol tetranitrate
PBX Polymer-bonded explosive
ppm Part per million
rem Roentgen equivalent in man
RDX Cyclotrimethylenetrinitramine
SA Arsine (blood agent)
SAIC Science Applications International Corporation
SFC Supercritical fluid chromatography
T bis(2-chloroethyl sulfide) monoxide
TATB 1,3,5-Triamino-2,4,6-trinitrobenzene
TEA Thermal energy analyzer
TLC Thin layer chromatography
TNT Trinitrotoluene
U.S. United States
U.S.S.R Union of Soviet Socialist Republics
UV Ultraviolet
VX V gas (nerve agent)

Glossary

Absorbed dose Energy imparted to matter by ionizing radiation per unit mass of irradiated material at the place of interest in that material; expressed in **rad** units.

Agroterrorism Form of terrorism that focuses on poisoning food supplies.

Alpha particle Positively charged (+2) radiation particle identical to the nucleus of a helium atom that consists of two protons and two neutrons.

al-Qaida Worldwide terrorist network associated with Osama bin Ladin.

Anthrax Bacterial agent that may cause inhalation anthrax, cutaneous anthrax, or gastrointestinal anthrax disease.

Atomic mass Number equal to the number of protons plus the number of neutrons present in the nucleus of an atom.

Atomic number Experimentally determined number characteristic of an element that is equal to the number of protons in the nucleus.

Beta particle Radiation in the form of an energetic electron emitted from the nucleus of an atom. Negatively charged (–1) radiation particle consisting of an energetic electron emitted from the nucleus of an atom.

Biological agent Bacteria, virus, or toxin designed for use as a weapon.

Blistering agent Agent that affects the eyes and lungs and blisters the skin; the five categories of blistering agents are mustards, arsenicals, nitrogen mustards, oximes, and mixes.

Blood agent Agent absorbed into the body primarily through breathing; affects the body by preventing normal utilization of oxygen by cells, which causes rapid damage to body tissue. The most common blood agents include hydrogen cyanide (AC), cyanogen chloride (CK), and arsine (SA).

Botulism Toxin that occurs in the form of food-borne botulism, wound botulism, and infant botulism; one of the most toxic substances known to man.

Brucellosis Disease caused by *Brucella melitensis, B. suis, B. abortus, or B. canis* bacteria species.

Chemical agent Chemical designed for use as a weapon; primary types include blister agents, blood agents, choking agents, and nerve agents.

Choking agents Chemical agents designed to target the respiratory tract; they cause lungs to fill with liquid, and death results from dry land drowning. The primary agents include chlorine, phosgene, and diphosgene.

Cholera Toxin that causes the mucosal cells of the small intestine to hypersecrete water and electrolytes into the lumen of the gastrointestinal tract.

Clostridium perfringens Pathogenic bacterium that causes toxin-mediated pulmonary syndrome.

Combustion Self-sustained, exothermic, rapid oxidation reaction of explosive substance and propellant.

Conventional explosive Explosive other than a nuclear weapon. Examples include nitroglycerine, picric acid, TNT, PETN, RDX, HMX, and TATB.

Copious Abundant, plentiful, present in large quantity.

Decontamination Removal of contamination from personnel or objects.

Deflagrating explosive A substance (e.g., propellant) that reacts by deflagration rather than detonation when ignited and used in a normal manner.

Deflagration Chemical reaction in which the output of heat is sufficient to enable the reaction to proceed and be accelerated without heat from another source; a surface phenomenon in which reaction products flow away from the unreacted material along the surface at subsonic velocity. The result of a true deflagration under confinement is an explosion.

Depleted uranium Uranium composed predominantly of the U-238 isotope; considered "depleted" because most of the U-235 component has been removed.

Detonating explosive Substance that reacts by detonation rather than deflagration when initiated and used in a normal manner.

Detonation Most violent type of explosive event; supersonic decomposition reaction that propagates through energetic material to produce an intense shock in the surrounding medium (air or water) and rapid plastic deformation.

Detonator Device for initiating an explosive that requires a shock wave. Initiation may be via electrical means, friction, flash from another igniferous element, stabbing, or percussion. A detonator may be constructed to detonate instantaneously or may contain a delay element.

Dirty bomb A conventional explosive device also containing radioactive materials used for the express purpose of contaminating property and/or terrorizing the general public.

Emergency responder Safety professional (fireman, police officer, medical technician, etc.) who will respond to a terrorist attack involving nuclear, chemical, or biological weapons.

Explosive train Arrangement of explosive components in which the initial force from the primer is transmitted and intensified until it reaches and sets off the main explosive composition.

Exposure Contact with ionizing radiation or radioactive material.

External dose Portion of dose equivalent received from radiation sources outside the body.

Fallout Radioactive material that falls back to earth after a nuclear explosion. Contains highly radioactive materials from the original weapon, vaporized material from ground zero, and other materials pulled into the mushroom cloud. The amount of fallout and spread of radioactivity depends on weapon yield and meteorological conditions.

Gamma particle Energetic photon (particle of light) originating from the nucleus of an atom produced when a neutron or proton drops from a high energy level to a lower energy level.

Glanders Disease that mainly affects horses, but can be fatal in man.

Gray (Gy) SI unit of absorbed dose; equal to an absorbed dose of 1 joule/kilogram (100 rads).

Ground zero Location of a nuclear explosion. Exact location may be in the air (using a bomb or missile) or underground but is assumed in this text to be detonated on the ground, where it will do the most damage.

Half-life Time required for half of a radioactive substance to disintegrate by radioactive decay.

Hemorrhagic fevers Ebola virus, Marburg virus, Lassa fever, Argentine and Bolivian hemorrhagic fevers, Crimean Congo hemorrhagic fever, Rift Valley fever, dengue hemorrhagic fever, and yellow fever.

High radiation area Accessible area where radiation levels could allow an individual to receive a dose equivalent in excess of 0.1 rem (1 mSv) in 1 hour at 30 centimeters from the radiation source or from any surface the radiation penetrates.

Highly enriched uranium Uranium comprised of >20% U-235.

Igniter Small device loaded with an explosive that will deflagrate; the output is primarily heat (flash), sometimes referred to as a squib.

Ignition Reaction occurring when a combustible material such as an explosive is heated to or above its ignition temperature.

Ignition means Method employed to ignite a deflagration train of explosions of pyrotechnic substances (e.g., a primer for propelling a charge, an igniter for a rocket motor, or an igniting fuse).

Ignition temperature Minimum temperature required for the process of initiation to be self-sustained.

Ingestion Process of taking material into the body via the digestive tract.

Inhalation Drawing air into the lungs through the nose and mouth.

Initiating device Another term for primary explosive.

Initiation Bringing an explosive to the state of deflagration or detonation.

Internal dose Portion of the dose equivalent received from radioactive material taken into the body.

Islamic Jihad Shiite organization derived from the Amal movement; acts through a number of subsidiary groups engaged in terrorism; Hizballah and the Islamic Amal are its most successful branches.

Isotope Any of two or more species of atoms of a chemical element with the same atomic number and different atomic mass. For example, U-238 and U-235 are both isotopes of uranium.

Lethal Capable of causing death.

LC_{50} Agent concentration in air that will kill 50% of those exposed through inhalation.

LD_{50} Amount of liquid or solid material that will kill 50% of those exposed through skin absorption or ingestion.

Melioidosis Bacterial disease of rodents that can be transmitted to humans via food contaminated by rodent droppings or biting flies.

mrem Millirem; one-thousandth of a rem. See **rem**.

Mushroom cloud Giant mushroom-shaped cloud extending from ground zero well into the atmosphere; initially contains an immense fireball and highly radioactive materials that eventually fall back to earth at various distances from ground zero.

Nerve agent Chemical agent inhaled, absorbed through the skin, or ingested that causes interference with the neural synapses and overstimulation of the nervous system, which in turn leads to overreactivity in the muscles and malfunctioning of various organs. The primary agents are tabun, sarin, soman, and VX.

Nihilistic terrorism Terrorism focusing on massive death and destruction of property.

Nuclear explosive device Assembly of nuclear and other materials and fuses that could be used in a test but generally cannot be delivered reliably as part of a weapon.

Nuclear fuel rod Fuel element in a nuclear reactor, typically composed predominantly of U-238 and U-235; could be used in a **dirty bomb**.

Nuclear warhead Refined and predictable nuclear device that can be carried by missile, aircraft, or other means.

Nuclear weapon Fully integrated nuclear warhead with delivery system.

Plague Bacterial agent that may cause bubonic plague or pneumonic plague.

Politically motivated terrorism Type of terrorism that limits the amount of violence to ensure that it does not impact supporters' sympathy for its cause.

Q fever Highly infectious disease caused by *Coxiella burnetii* bacteria; a single bacterial cell can produce clinical illness.

Rad Special unit of absorbed dose; one rad equals an absorbed dose of 100 ergs/gram or 0.01 joule/kilogram (0.01 Gy).

Radiation (ionizing) Alpha particles, beta particles, gamma rays, x-rays, and other particles capable of producing ions; does not include nonionizing radiation forms such as radio waves, microwaves, or visible, infrared, or ultraviolet light.

Radiation area Accessible area where radiation levels could allow an individual to receive a dose equivalent in excess of 0.005 rem (0.05 mSv) in 1 hour at 30 centimeters from the radiation source or from any surface the radiation penetrates.

Radiation dose Level of exposure to radiation, expressed in units called Roentgen equivalents in man (rem); a dose of approximately 400 rem is fatal in 50% of cases when medical treatment is not available. No health effects are typically observed at doses below 10 rem.

Radioactivity Property or characteristic of radioactive material to spontaneously disintegrate with the emission of energy in the form of radiation; measured in curies or becquerel.

Rem (Roentgen equivalent in man) Special unit of any quantity expressed as a dose equivalent; the dose equivalent in rems is equal to the absorbed dose in rads multiplied by the quality factor (1 rem = 0.01 Sv).

Ricin Toxin that poisons the body.

Rift Valley fever Viral fever found primarily in sub-Saharan Africa where it is transmitted by mosquitoes.

Saxitoxins Water-soluble toxins that prevent proper nerve functioning.

Sievert SI unit of any of a quantity expressed as a dose equivalent; the dose equivalent in sieverts is equal to the absorbed dose in grays multiplied by the quality factor (1 Sv = 100 rems).

Smallpox Virus transmissible via large or small respiratory droplets or contact with skin lesions and secretions.

Staphylococcus One of the most toxin-producing germs known.

Strategic nuclear weapon Long-range nuclear weapon generally allocated for attacking an enemy or protecting the homeland.

Tactical nuclear weapon Nuclear weapon intended to affect the outcome of a tactical maneuver or battle.

Trichothecene mycotoxin Toxin produced by fungal molds; it inhibits protein synthesis, impairs DNA synthesis, and interferes with cell membrane structure and function.

Tularemia Disease caused by a bacterial agent. Examples include ulceroglandular tularemia, glandular tularemia, typhoidal tularemia, oculoglandular tularemia, oropharyngeal tularemia, and pneumonic tularemia.

Venezuelan equine encephalitis Influenza-like disease caused by a virus leading to neurologic complications.

Very high radiation area Accessible area where radiation levels could allow an individual to receive an absorbed dose in excess of 500 rads (5 Gy) in 1 hour at 1 meter from a radiation source or from any surface the radiation penetrates.

Weapon of mass destruction Weapon that has the potential of inflicting a mass number of casualties. While the phrase has been used historically to refer to nuclear weapons, it also includes weapons that distribute chemical and/or biological agents.

Yield Strength of a nuclear weapon, usually expressed in tons of TNT. The bomb dropped on Hiroshima, Japan in 1945 had a yield of approximately 15,000 tons (15 kilotons) of TNT. Modern high-yield weapons have yields >1000 kilotons of TNT. The 4000 lb ammonium nitrate bomb used to blow up the Oklahoma City Federal Building in 1995 was equivalent to about 1.5 tons (0.0015 kilotons) of TNT.

1 Introduction

1.1 PURPOSE AND SCOPE

Following the events of September 11, 2001, the United States began a strategic assault against terrorist groups around the world with the objective of ridding the world of large terrorist organizations that could potentially repeat a September 11 type event.

In an effort to protect the American public from future large-scale terrorist activities on U.S. soil, President George W. Bush proposed the most extensive reorganization of the federal government since the 1940s by creating a new Department of Homeland Security. The primary mission of this new department is to:

- Reduce America's vulnerability to terrorism
- Coordinate all efforts to secure the American people against bioterrorism and other weapons of mass destruction
- Minimize the damage from terrorist attacks
- Recover from terrorist attacks
- Train and equip first responders
- Manage federal emergency response activities

The Department of Homeland Security is designed to be comprised of four main divisions: Border and Transportation Security; Emergency Preparedness and Response; Chemical, Biological, Radiological and Nuclear Countermeasures; and Information Analysis and Infrastructure Protection. In addition to the responsibilities described above, the new department is intended to:

- Set national policy and establish guidelines for state and local governments
- Direct exercises for federal, state, and local chemical, biological, radiological, and nuclear attack response teams and plans
- Consolidate and synchronize the efforts of multiple federal agencies now scattered across several departments
- Improve America's ability to develop diagnostics, vaccines, antibodies, antidotes, and other countermeasures against new weapons
- Assist state and local public safety agencies by evaluating equipment and setting standards

While the establishment of a Department of Homeland Security is a good first step in helping protect the American people from future September 11 type terrorist attacks, it brings no guarantees. The American public has been forced to come to grips with the strong possibility of future large-scale terrorist events occurring within their borders. The public has also come to realize that they play a critical role in helping prevent these types of events. This new "take-charge" attitude was clearly demonstrated on December 22, 2001, when passengers and flight crew members on American Airlines Flight 63 (en route from Paris to Miami) tackled Richard C. Reid when he attempted to set off a bomb hidden inside his footwear.

All American citizens can play active roles in preventing future terrorist events by reporting suspicious activities to their local police departments and acting when necessary as demonstrated by the passengers on American Airlines Flight 63. While the public can gain valuable information from this book on protection from terrorist activities involving nuclear, chemical, and biological weapons (weapons of mass destruction), the target audiences for this book are safety professionals, law enforcement officials, and Federal Bureau of Investigation agents (collectively referred to as emergency responders) since they have an ever-increasing likelihood of receiving first-hand exposure to one or more terrorist events involving these types of weapons. The primary objective of this book is to provide emergency responders with guidance on:

- Types of weapons of mass destruction that could be used in a terrorist attack
- Mechanisms by which terrorists could disperse various types of nuclear, chemical, or biological agents
- Types of conventional explosives that terrorists could use to disperse these agents
- Routes by which one may be exposed to these agents
- Types of health hazards that may result from exposure to these agents
- Steps for minimizing exposure of safety professionals to these agents
- Potential medical treatment options for those receiving exposure to these agents
- Best methods of increasing chances of surviving a nuclear explosion
- Emergency preparedness measures for a variety of settings
- How to prioritize injuries
- Personnel decontamination procedures to be implemented prior to medical treatment
- Radiation exposure guidelines
- Training guidelines

The following sections provide a historical perspective on terrorist activities targeted against the U.S., along with details on the historical development of weapons of mass destruction.

1.2 HISTORICAL PERSPECTIVE ON TERRORISM TARGETED AGAINST THE U.S.

Studies performed several years prior to September 11, 2001 showed that terrorist incidents around the world declined somewhat, while the severity of individual attacks increased dramatically.[1] The motivation for terrorist acts has shifted in recent years from being politically driven to having a more fatalistic orientation. Current-day terrorism is increasingly motivated by body count, and more often than not is religiously or ethnically motivated. While politically motivated terrorism tends to limit the amount of violence in order to ensure that it does not impact supporters' sympathy for its cause, today's nihilistic terrorism is more simply focused on massive death and destruction of property.

This suggests that terrorists are driven to use bigger bombs or more deadly weapons. Based on this premise, it seems very likely that in the near future terrorists will elect to use more weapons of mass destruction rather than traditional explosives or firearms. Biological weapons are particularly likely to be used by terrorist groups because they:

- Are highly lethal
- Require only modest size relative to their destructive potential
- Can easily be used without explosives (e.g., contaminating drinking water supplies)
- Are relatively simple and able to be rapidly deployed
- Are relatively inexpensive as compared to nuclear weapons

Biological agents also provide terrorists with a wide variety of alternatives for dispersion. For example, rather than using explosives to disperse biological agents, these materials could potentially:

- Contaminate food or water supplies
- Be dispersed as vapors or by aerosols within an enclosed area (e.g., building, tunnel) where the ventilation system would further distribute the agent
- Be dispersed as vapors or by aerosols in an open area where the wind would carry the agent great distances
- Be transmitted through infected animals or insects (e.g., fleas, ticks, flies, rats)

The U.S. Office of Technology Assessment notes that an aerosol attack on the Washington, D.C. area could yield as many as three million casualties. Other sources suggest approximately 500,000 casualties from a less ambitious attack.

The historical Cold War strategy for avoiding encounters with weapons of mass destruction consisted almost entirely of deterrence through arms control, intelligence collection, diplomacy, and threats of armed force. These methods of deterrence are no longer as effective because religiously motivated terrorists believe they are carrying out the will of God. Since terrorist groups are not often tied to any specific country, the U.S. often has no clear target for its retaliation. As a result, terrorist groups feel they can launch attacks on the U.S. and escape punishment. The following is a chronology of the more significant historical terrorist attacks against the U.S. within its borders and abroad.

1972: Typhoid Bacteria Food Poisoning — In 1972, members of a U.S. fascist group called Order of the Rising Sun were found in possession of 30 to 40 kilograms of typhoid bacteria cultures they planned to use to contaminate water supplies in Chicago, St. Louis, and other large Midwestern cities.

1975: LaGuardia Airport Bombing — The Armed Forces for National Liberation claimed responsibility for the December 29, 1975 bombing of a Transworld Airlines terminal at LaGuardia Airport in New York that killed 11 people.

1979: Seizure of the U.S. Embassy in Tehran — The U.S. Embassy in Tehran was overtaken by Iranian revolutionaries and students on November 4, 1979. One disastrous rescue (Desert One) was attempted on April 24, 1980 under the direction of President Jimmy Carter. This mission was designed to use a team of helicopters based on carriers in the Indian Ocean to stage a raid on the compound where the hostages were held. Rough seas and a fierce sandstorm in the desert combined to down three of the eight helicopters. Another helicopter collided with a C-130 transport aircraft in the desert. The rescue mission had to be aborted. All 52 American hostages were released on January 20, 1981, after they were held in captivity for 444 days. President Ronald Reagan was in charge at the time of the release, which was made in return for unfreezing Iranian assets held in U.S. banks and supplying U.S. weapons and spare parts.

1983: Bombing of U.S. Marine Barracks, Beirut — On April 18, 1983, a vehicle driven by a member of the Islamic Jihad crashed into the outer wall of the U.S. Embassy in Beirut, killing 67 people, 17 of whom comprised the majority of Central Intelligence Agency (CIA) staff in Lebanon. In retaliation, the U.S. Navy bombarded terrorist positions designed to aid the Lebanese army in Beirut. In response, on October 23, 1983, the Islamic Jihad fitted two trucks with 12,000 pounds of explosives and sent them on a suicide mission. The trucks passed the lax security at the U.S. Marine guardhouse and crashed into a wall of a four-story concrete barrack. All 241 marines and 58 French soldiers inside were killed in the explosion.

1984: Kidnapping and Murder of William Buckley — William Buckley, a U.S. Embassy diplomat and CIA station chief in Beirut, was kidnapped on March 16, 1984 by Islamic Jihad gunmen in Muslim west Beirut. Buckley was reportedly secretly transported through Syria by Iranian gunmen and delivered to Iran for interrogation. Fearful of the consequences if terrorists gained information from Buckley, the U.S. made extensive efforts to find and rescue him, but the attempts were not successful. Before he was killed, Buckley was forced to provide information that cost the lives of many undercover agents and sources. His decomposed body was dumped in southern Beirut nearly 8 years later.

1984: American Embassy Complex Bombing, Beirut — On September 20, 1984, fanatical members of the Islamic Jihad drove a truck loaded with explosives into the U.S. embassy complex. A steel gate designed to protect the annex was lying on the ground awaiting installation. The driver of the truck was shot by the bodyguard of British ambassador David Miers, causing the truck to swerve and miss the annex.

Nine people were killed in the explosion, but the annex full of American citizens was saved from destruction.

1985: Attempted Missile Purchase, Chicago — During 1985, a terrorist group (El Ruken) linked to Libya used violence and intimidation on the streets of Chicago to raise money to purchase weapons. Several members of the group were arrested in 1985 when they attempted to purchase a hand-held missile to shoot down an airplane at Chicago's O'Hare Airport.

1985: Hijacking of TWA Flight 847 — On June 14, 1985, a Hizballah terrorist group under the leadership of Mohammed Ali Hamadi hijacked TWA Flight 847 after it left Athens, Greece, and ordered the plane to Beirut. Hamadi held 39 passengers hostage for 17 days. When Hamadi's demands were not met, Petty Officer Stethem was singled out from the passengers as a U.S. Navy sailor and was tortured and killed before his body was dumped onto the tarmac.

1986: La Belle Discotheque Bombing, Berlin — The La Belle Discotheque in Berlin, which was frequently attended by American soldiers and personnel, was bombed on April 5, 1986, and 41 of the 234 persons injured were Americans. Two U.S. soldiers died in the attack. Experts concluded that a military explosive (e.g., Semtex) was used in the attack and that Libya took part in the planning and execution of the bombing. The U.S. retaliated with concentrated air strikes on targets in Libya. The air strikes apparently led Libya's Colonel Qaddafi to finance the plot to bring down Pan Am Flight 103.

1988: Plan to Bomb Manhattan Office Buildings — On April 12, 1988, a New Jersey state trooper noticed a man acting strangely at a rest stop on the New Jersey Turnpike. The trooper stopped the man, and upon further investigation discovered three large bombs in the trunk of the man's car. Yu Kikumura, the driver, admitted he was a member of the Japanese Red Army and that the bombs were intended for midtown Manhattan office buildings. Kikumura was suspected of working for Libya.

1988: Bombing of Pan American Flight 103 — Pan American Flight 103 exploded over Lockerbie, Scotland, on December 21, 1988, resulting in 270 American and Scottish fatalities. The explosion was caused by a barometrically triggered bomb concealed in a suitcase. The men responsible are believed to be hiding in Libya.

1991: El Salvador Attack — On January 2, 1991, Farabundo Marti National Liberation Front militants in San Miguel downed a U.S. helicopter and executed two U.S. crewmen (Lieutenant Colonel David Pickett and Crew Chief Earnest Dawson). A third American (Chief Warrant Officer Daniel Scott) died of injuries received in the crash.

1991: Athens Attack — Air Force Sergeant Ronald Steward was killed on March 12, 1991 by a remote-controlled bomb detonated at the entrance of his apartment building in Athens, Greece. A revolutionary organization claimed the attack was in response to "the genocide of 13,000 Iraqis."

1991: Islamic Jihad Bombings in Turkey — Two car bombings occurred in Turkey on October 28, 1991 and killed one Air Force sergeant and severely injured an Egyptian diplomat. The Turkish Islamic Jihad claimed responsibility.

1991: Beirut Attack — On October 29, 1991, a rocket struck the edge of the U.S. Embassy in Beirut, Lebanon. There were no casualties.

1991: Beirut Bombing — On November 8, 1991, a 100-kilogram car bomb destroyed the administration building of the American University. The attack killed one person and wounded at least a dozen others.

1993: World Trade Center Bombing — At 12:18 p.m. on February 26, 1993, an improvised explosive device exploded on the second level of the World Trade Center parking basement. The resulting blast produced a crater approximately 150 feet in diameter and 5 floors deep. The structure consisted mainly of steel-reinforced concrete, 12 to 14 inches thick. The epicenter of the blast was approximately 8 feet from the south wall of Tower One. The device was placed in the rear cargo portion of a 1-ton Ford F350 Econoline van rented from the Ryder agency in Jersey City, New Jersey. Approximately 6800 tons of material were displaced by the blast. The explosion killed six people and injured more than a thousand. More than 50,000 people were evacuated from the complex during the hours immediately following the blast.

The main explosive charge consisted primarily of 1200 to 1500 pounds of urea nitrate, a home-made fertilizer-based explosive. The bomb was packed with cyanide with the intent of spreading the poison throughout the building to make it uninhabitable, but the method was ineffective. The fusing system consisted of two 20-minute lengths of a nonelectric burning fuse. Also incorporated in the device and placed under the main explosive charge were three large metal cylinders (126 pounds) of compressed hydrogen gas.

On March 3, 1993, a typewritten communication received at *The New York Times* noted that the World Trade Center was bombed in the name of Allah. Mohammad Salameh, Nidel Ayyad, Ahmad Ajaj, and Mahmud Abouhalima were all identified as suspects and later found guilty after a 6-month trial.

1994: Oregon Cult Food Poisoning — Two members of an Oregon cult headed by Bhagwan Shree Rajneesh cultivated Salmonella (food poisoning) bacteria and used them to contaminate restaurant salad bars in an attempt to affect the outcome of a local election. Although hundreds of people became ill and 45 were hospitalized, there were no fatalities.

1995: Aryan Nations Member Caught Ordering Freeze-Dried Bacteria — A member of the Aryan Nations neo-Nazi organization was arrested in Ohio on charges of mail and wire fraud. He allegedly misrepresented himself when ordering three vials of freeze-dried *Yersinia pestis*, a bacterium that causes plague, from a Maryland biological laboratory.

1995: Arkansas Resident Charged with Possession of Ricin — Canadian customs officials intercepted a man carrying a stack of currency. A white powder was interspersed between the bills. Suspecting cocaine, customs personnel had the material analyzed and discovered that it was ricin, a strong toxin, and not cocaine. The Arkansas resident was charged with possession of ricin in violation of the Biological Weapons Anti-Terrorism Act of 1989.

1995: Minnesota Patriots Council Planned Use of Ricin — In March 1995, two members of the Minnesota Patriots Council, a right-wing militia organization advocating a violent overthrow of the U.S. government, were convicted of conspiracy charges for planning to use ricin, a biological toxin.

1995: Planned Sarin Attack at Disneyland, California — In April 1995, *The Baltimore Sun* reported that a planned gas attack on Disneyland had been uncovered. Two Japanese men affiliated with Aum Shinri Kyo were in possession of instructions for making sarin and a videotape detailing an attack on Disneyland; they were arrested at Los Angeles International Airport.

1995: Oklahoma City Bombing — On April 19, 1995 at around 9:03 a.m., a massive bomb exploded inside a rental truck, destroying half of the 9-story Alfred P. Murrah Federal Building in downtown Oklahoma City. A total of 168 people were killed. The components of the bomb included nitromethane, ammonium nitrate fertilizer, Tovex sausage explosives, shock tube, and cannon fuse. Only 90 minutes after the explosion, an Oklahoma Highway Patrol officer stopped 27-year-old Timothy McVeigh for driving without a license plate. He was almost released on April 21, before he was recognized as a bombing suspect. McVeigh and his ex-Army partner, Terry Nichols, were charged and later convicted.

1998: Bombing of U.S. Embassies in Nairobi and Dar-es-Salaam — On August 7, 1998, 224 innocent civilians were killed and over 5000 were wounded by terrorist bombs exploded at the U.S. embassies in Nairobi, Kenya and Dar es Salaam, Tanzania. The terrorists responsible are believed to be part of an international criminal conspiracy headed by Osama bin Laden.

2000: Bombing of U.S.S. Cole — On October 12, 2000, a small boat assisting in the refueling of the U.S.S. Cole, a destroyer, at the Port of Aden, Yemen, exploded and killed 17 sailors. The small boat was packed with sophisticated explosives that tore a 20- by 40-foot gash in the midhull of the ship. The attackers are believed to be tied to Osama bin Ladin and his worldwide al-Qaida network.

2001: Attack on World Trade Center and Pentagon — On September 11, 2001, two American Airlines and two United Airlines planes were hijacked and later used as terrorist weapons. Two of the hijacked planes crashed into the World Trade Center towers, while a third airline crashed into the Pentagon in Washington. The fourth plane crashed in Somerset County, Pennsylvania before reaching its target, which may have been the White House. The prime terrorist suspects behind the attack are Osama bin Laden and his al-Qaida network.

The attack resulted in the loss of approximately 184 lives at the Pentagon and nearly 3000 in New York and Pennsylvania. Approximately 42 of those who died were citizens of other counties. Both World Trade Center towers were completely destroyed and the Pentagon suffered large-scale damage. A chronology of the September 11 events is as follows.

8:45 a.m.: A hijacked passenger jet, American Airlines Flight 11 from Boston crashes into the north tower of the World Trade Center, tearing a hole in the building and setting it on fire.

9:03 a.m.: A second hijacked airliner, United Airlines Flight 175 from Boston crashes into the south tower and explodes.

9:17 a.m.: The Federal Aviation Administration shuts down all New York City area airports.

9:21 a.m.: The Port Authority of New York and New Jersey orders all bridges and tunnels in the area closed.

9:30 a.m.: President Bush, speaking in Sarasota, Florida, says the country suffered an "apparent terrorist attack."
9:40 a.m.: The Federal Aviation Administration halts all flight operations at U.S. airports, the first time in U.S. history that nationwide air traffic is halted.
9:43 a.m.: American Airlines Flight 77 crashes into the Pentagon, sending up a huge plume of smoke. Evacuation begins immediately.
9:45 a.m.: The White House evacuates.
9:57 a.m.: President Bush leaves Florida.
10:05 a.m.: The south tower of the World Trade Center collapses, plummeting into the streets below. A massive cloud of dust and debris forms and slowly drifts away from the building.
10:08 a.m.: U.S. Secret Service agents armed with automatic rifles are deployed into Lafayette Park across from the White House.
10:10 a.m.: A portion of the Pentagon collapses.
10:10 a.m.: Hijacked United Airlines Flight 93 crashes in Somerset County, Pennsylvania, southeast of Pittsburgh.

President Bush responded to the attacks by ousting the Taliban from power in Afghanistan after the Taliban refused to turn over Osama bin Laden to face trial for the September 11 terrorist attacks.

2001: Anthrax Attacks — Soon after September 11, 2001, terrorists began sending envelopes containing anthrax spores through the U.S. mail system. The envelopes were mailed to NBC and CBS television stations, *The New York Post*, Senator Tom Daschle's office, the State Department, and other locations. As of November 21, 2001, 37 individuals had been exposed to anthrax; 13 were infected and 5 eventually died. The FBI is still trying to determine who was responsible for these attacks.

2001: Shoe Bomber — Richard C. Reid is currently awaiting trial in the U.S. for allegedly trying to set off a bomb hidden in his footwear during a flight from Paris to Miami on December 22, 2001. Two suspects from Pakistan and three from North Africa are also being detained for possible involvement.

2002: Kuwaiti Gunmen Attack U.S. Forces — On October 8, 2002, two Kuwaiti gunmen in a pickup truck attacked U.S. forces during war games on an island in the Persian Gulf. One Marine was killed and one was wounded before the gunmen were shot to death by U.S. troops. The Kuwaiti assailants were identified as Anas al-Kandari, born in 1981, and Jassem al-Hajiri, born in 1976. U.S. intelligence has not yet determined whether they had terrorist links.

2002: Bombing in Bali — A bomb was detonated at a nightclub in Bali, Indonesia, in October 2002. The blast killed nearly 200 people, including 2 Americans, 15 Australians, 3 Singaporeans, 2 Britons, several others from other European countries, and one Ecuadorean. Three other Americans were wounded. This was a different type of terrorism from what the Bush administration campaigned against because the target was not an American embassy, military outpost, or financial institution representing American power. Rather, it was a nightclub frequented mostly by Europeans and Australians. The U.S. has not yet determined who was responsible.

2002: Beltway Snipers — Thirteen people were shot, 10 fatally, in Maryland, Virginia, and Washington, D.C. in a series of related sniper attacks that began October 2. On October 24, authorities arrested John Allen Muhammad and John Lee Malvo who were identified as suspects. In addition, police seized a gun used in 11 of the shootings. Whether the shootings were terrorist or criminal acts has not been determined.

2002: Killing of U.S. Diplomat in Amman — As he walked to his car in front of his home in Amman, Jordan, American diplomat Laurence Foley was shot at least seven times in the head and chest by a lone gunman at close range. Foley was an administrator at the U.S. Agency for International Development. The gunman escaped. It is believed that the attack may have been coordinated by al-Qaida or its sympathizers.

1.3 HISTORICAL DEVELOPMENT OF NUCLEAR WEAPONS

The building blocks that supported the development of the first nuclear weapon were developed in 1789 when Martin Heinrick Klaproth discovered uranium. While its original use was as a glass coloring agent, later researchers discovered more complex uses. A few of the more prominent scientists who helped develop our current understanding of radiation theory and/or assisted in the development of the first atom bomb include Henri Becquerel, Marie Curie, Ernest Rutherford, Albert Einstein, Otto Frisch, Enrico Fermi, Leo Szilard, Samuel Allison, Glenn Seaborg, Arthur Compton, and Tokutaro Haiwara.

This section summarizes the key events that led the United States to be the first country to successfully develop the atomic bomb. It also discusses the world's current nuclear weapons arsenal and details the uses of other types of radioactive materials as terrorist weapons.

1.3.1 The Race to Develop the First Nuclear Weapon

On October 11, 1939, a letter signed by Albert Einstein was delivered to President Franklin D. Roosevelt. The letter warned of the potential for and consequences of atomic weapons and suggested that the president take immediate action to counter the work in progress at the Kaiser Wilhelm Institute of Berlin. Since the Presidential Advisory Committee continued to focus on uranium research as opposed to application, it was replaced in June 1940 by the National Defense Research Committee led by Vannevar Bush, a scientist who introduced security and secrecy into nuclear research by barring foreign-born scientists from the committee and blocking the publication of articles on uranium research. In November 1940, Enrico Fermi and Leo Szilard began constructing a subcritical, graphite-moderated, uranium oxide reactor at Columbia University to further investigate chain reactions.

On June 28, 1941, President Roosevelt established the Office of Scientific Research and Development. Vannevar Bush was named its director and reported directly to the President. This new office was created within a week of Germany's invasion of the Soviet Union and less than a month after Tokutaro Haiwara gave a presentation at the University of Kyoto in which he speculated about the potential for a fusion explosion using fission ignition which serves as the basis for thermonuclear weapons.[2]

The British issued a report to the U.S. on July 15, 1941 and described the technical details of an atomic bomb along with proposals and cost estimates for its development. After the report was brought to President Roosevelt's attention, he accelerated the pace of research to determine the feasibility of a bomb. On November 6, 1941, Arthur Compton estimated that a critical mass of 2 to 100 kilograms of uranium-235 would produce a powerful fission bomb and would cost an estimated $50 million to $100 million. Vannevar Bush provided this information to the President on November 27, 1941 and was authorized to organize an accelerated research project to investigate gaseous diffusion, electromagnetic separation, centrifuge separation, chain reaction, heavy water production, and plutonium production. Bush began constructing pilot plants.

Shortly after Japan's December 7, 1941 attack on Pearl Harbor, the U.S. became more driven to expedite its timetable for developing the first fission weapon because of fear that the U.S. lagged behind Nazi Germany in efforts to create the first atomic bomb. On December 2, 1942 at 3:49 p.m., Enrico Fermi and Samuel K. Allison achieved the world's first controlled, self-sustained nuclear chain reaction in an experimental reactor using natural uranium and graphite.

Concurrent with the search for technologies to produce isotopes of uranium-235 (U-235) and plutonium-239 (Pu-239) was the search for a method to separate the isotopes once they were produced. In order to determine whether U-235 would support an explosive nuclear chain reaction, concentrated samples of sufficient size to form a critical mass were required. Since the separation could not be accomplished through chemical means (U-238 and U-235 are chemically indistinguishable), physical separation by atomic weight was required. Electromagnetic separation, centrifuge separation, and gaseous diffusion techniques were investigated. Given the uncertainty that any of these separation methods would concentrate U-235 and the small likelihood of sustaining a nuclear reaction that would transform uranium into plutonium, the decision was made to proceed with all options simultaneously regardless of cost.

Later research by Glenn Seaborg and Emillio Segre revealed that Pu-239 was 1.7 times as fissionable as U-235, and was thus a better nuclear explosive. On August 20, 1942, Seaborg identified a sequence of chemical oxidation and reduction cycles that produced a microgram of plutonium.[3] On June 18, 1942, Brigadier General Wilhelm D. Styer established a U.S. Army Corps of Engineers District devoted exclusively to managing and coordinating atomic weapons development. This organization was named the Manhattan Engineer District, later designated the Manhattan Project.

On November 16, 1942, Los Alamos, New Mexico, was selected as the central site (Site Y) for a laboratory to research the physics and design of atomic weapons. Site X was at Oak Ridge, Tennessee and consisted of an experimental reactor, chemical separation plant, and electromagnetic separation facility. An area near

Hanford and Richland, Washington, was selected for industrial-scale plutonium production and chemical separations facilities on January 16, 1943. This site was named the Hanford Engineer Works (later named the Hanford Site).

In only 30 months, the Manhattan Project built 554 buildings including reactors, separation plants, laboratories, craft shops, warehouses, and electrical substations. The Hanford Site plutonium production reactors (B, D, and F) were rectangular, measured 36 feet long by 28 feet wide by 36 feet high, used 200 tons of uranium metal fuel and 1200 tons of graphite, were water cooled, and operated at an initial power level of 250 million watts (thermal). They dwarfed the reactors at other sites.

On September 13, 1944, the Hanford Site started the B Reactor. For approximately 1 hour all went well, but the reactor malfunctioned as a result of fission product poisons. On December 17, 1944, the Hanford Site D reactor was started and the B reactor was repaired and restarted. Large-scale plutonium production was under way. On February 25, 1945, the Hanford F Reactor was started. With these three reactors operating simultaneously, the theoretical plutonium production capacity was approximately 21 kilograms per month.

Plutonium from the Hanford Site was shipped to Los Alamos every 5 days, and enriched uranium was shipped to Los Alamos from Oak Ridge. At 5:30 a.m. on Monday, July 16, 1945, the U.S. tested the first plutonium bomb, named Trinity, at the White Sands Missile Range, New Mexico. The bomb exploded with a force of approximately 18.6 kilotons. After this test there was no longer any question that the plutonium bomb would work.

President Roosevelt did not live to see the atomic age. He died of a cerebral hemorrhage on April 12, 1945. Before President Harry S. Truman was fully advised of the Manhattan Project, a bombing raid on Tokyo destroyed the building containing Japan's gaseous thermal diffusion experiment, which ended Japan's atomic bomb project. About the same time, American forces confiscated Belgian uranium ore stored in Strassfurt, Germany, and that crippled the German atomic weapons program.

Despite Germany's surrender, Japan continued to resist the unconditional surrender demanded by the Allied Forces. Knowing that the U.S. would shortly have enriched uranium and plutonium bombs ready for use enabled Truman to avoid extending Japan an offer of surrender that allowed the Emperor to continue to rule. On July 26, 1945, the Potsdam Declaration was issued via radio to Japan. President Truman, Chiang Kai-Shek of Nationalist China, and Winston Churchill of Great Britain called on the Japanese government to "proclaim now the unconditional surrender of all Japanese armed forces. The alternative for Japan is prompt and utter destruction."[4] Japanese leadership rejected the declaration on July 29, 1945.

At 2:45 a.m. on the morning of August 6, 1945, the *Enola Gay* (code named Dimples 82) began its takeoff run carrying Little Boy, an enriched uranium gun-style bomb that had never been tested. Little Boy was dropped on Hiroshima, Japan and exploded at 8:16 a.m. It had a yield of approximately 15,000 tons (15 kilotons) of trinitrotoluene (TNT). In a radio release, Truman noted that he planned to drop additional atomic bombs if Japan did not offer its unconditional surrender. Japan refused and the Fat Man bomb (named in honor of Winston Churchill), containing Hanford Site plutonium, was dropped on Nagasaki. The original intended target for

Fat Man was the Kokura Arsenal on Kyushu Island, but poor weather conditions led to the bombing of the secondary target of Nagasaki. This bomb exploded 1650 feet above the slopes of the city with a force of 21,000 tons (21 kilotons) of TNT.

The Little Boy bomb initially killed approximately 70,000 people and injured another 70,000. By the end of 1945, the death toll rose to 140,000 due to radiation sickness. Five years later, the death toll was 200,000. The initial death rate from Fat Man was 40,000 with 60,000 injured; the death rate eventually rose to about 140,000.

1.3.2 Modern Nuclear Weapons

Since the end of World War II, the U.S. manufactured over 70,000 nuclear weapons (many of which have been retired) using 65 different designs to deter and, if necessary, fight a nuclear war. As of 2002, the U.S. nuclear stockpile consists of 7982 deployed nuclear weapons and 2700 contingency-stockpiled nuclear weapons for a total of 10,682.[7] This is a dramatic reduction from 1967 when the U.S. had a peak number of 32,000 stockpiled nuclear weapons in its arsenal.[6] It is believed that Russia currently has a similar number of strategic nuclear weapons. Other countries known to possess nuclear weapons of various sizes include United Kingdom, France, China, Pakistan, and India. Countries suspected of potentially having nuclear weapons include Israel, North Korea, Libya, Iraq, and Iran. It is estimated that China currently has approximately 20 nuclear missiles capable of reaching the U.S.

Today's nuclear weapons vary in both size and method of delivery. While the Hiroshima enriched uranium bomb delivered a yield of 15 kilotons, it was very small compared to the large (1000-kiloton or 1-megaton) bombs currently representative of the world's arsenal. The common types of nuclear weapons available today include a large variety of land-based missiles, sea-based missiles, tactical air launch missiles, bombs to be dropped from strategic bombers, and suitcase bombs. Common components of nuclear weapons created after the 1980s include fissile and/or fusion materials; sequencing microprocessors; chemical high explosives; neutron activators; arming systems (components that serve to ready [prearm], safe, or resafe [disarm]); firing systems; radar, pressure-sensitive, and time-sensitive fusing systems; and safety devices. Today's nuclear weapons often include a number of safety and control systems to minimize the chances of accidental, unauthorized, or inadvertent use; for example:

- The "two man rule" requiring a minimum of two authorized personnel present whenever people come in contact with nuclear weapons
- Code-controlled arming and fusing systems
- Unique electronic signal generators for arming and fusing systems
- Handprint electronic access readers
- Devices to prevent against accidental surges and firing

While it is very unlikely that terrorist groups could gain access to a U.S. nuclear weapon, there is a great concern that they could obtain smaller or more primitive nuclear weapons from other countries, particularly Russia, Pakistan, North Korea,

Libya, Iraq, or Iran. Former Russian Security Council Secretary Aleksandr Lebed stirred controversies in both Russia and the U.S. by alleging that the Russian government cannot account for 84 small atomic demolition munitions (suitcase bombs) manufactured in the U.S.S.R. during the Cold War. Lebed originally made the allegations in a closed meeting with a U.S. congressional delegation in May 1997. He informed the delegation that he could confirm the production of 132 suitcase bombs, but could only account for 48. When asked about the whereabouts of the other 84, Lebed replied "I have no idea."

His charges generated public controversy 3 months later when he repeated them in an interview with the CBS television news magazine *60 Minutes* broadcast on September 7, 1997. Other Russian officials initially dismissed Lebed's charges, saying all the country's nuclear weapons were accounted for and under strict control. Top-ranking Russian defense officials later went further and denied that the U.S.S.R. ever built such weapons, claiming they were too expensive to maintain and too heavy for practical use.

Lebed stood by his statement, and his charges were backed by Aleksey Yablokov, a former advisor to President Yeltsin, who told a U.S. Congressional subcommittee on October 2, 1997 that he was "absolutely sure" that such suitcase nuclear bombs were ordered in the 1970s by the Komitet Gosudarstvennoi Bezopaznosti (KGB). This controversy is not the first public discussion of whether former Soviet suitcase nuclear bombs are under adequate control in Russia. During 1995, a flurry of Russian media reports claimed that Chechen separatist fighters obtained such weapons.

If al-Qaida or another terrorist group gained control of one or more Russian suitcase nuclear weapons, they could be smuggled into the U.S. by small boat or overland from Mexico or Canada. The explosion of such a device in a crowded city could cause immediate deaths of tens of thousands and lead to cancer for many of the survivors.

Terrorist access to Pakistan's 20 to 50 nuclear weapons is not out of the realm of possibility. Pakistan was deeply involved in Afghanistan in the 1979 to 1989 war against the U.S.S.R. The U.S. channeled weapons, training, and money through Pakistan to the Mujahadeen, the Taliban, and al-Qaida to fight the U.S.S.R. After September 11, 2001, the U.S. switched sides and requested Pakistan join the U.S. in fighting its former allies. The U.S. desperately needed Pakistan's airports to fight a war in Afghanistan. Many Pakistani supporters of al-Qaida and the Taliban were not behind the U.S. These supporters include members of Pakistani military, who could decide to rebel and gain control of some or all of Pakistan's nuclear arsenal.

1.3.3 Modern Weapons Using Depleted-Uranium Projectiles

In 1978, the U.S. Department of Defense began manufacturing military ammunition using depleted U-238, since it had more than 700,000 tons of this byproduct material left from nuclear weapon and nuclear power production. The material was attractive for ammunition production since it had no other use, cost nothing to produce, and is pyrophoric (bursts into flames on contact with a target).

Depleted uranium is composed mostly of the U-238 isotope and is considered "depleted" because most of its U-235 component has been removed. Natural uranium is composed of approximately 0.7% U-235 and 99.3% U-238. Depleted uranium has only about half the radioactivity of the original natural element, but radiation emanating from depleted uranium can be hazardous to human health and the environment.

Depleted uranium is 1.7 times more dense than lead and extremely effective for penetrating metal armor. For example, a 120-millimeter tank round contains about 10 pounds of solid uranium. At high speed it can easily slice through tank armor. The U.S. also uses U-238 as armor plating to prevent penetration by conventional weapons. When a depleted uranium projectile strikes a target, as much as 70% of the round is vaporized and converted to small particles of oxidized U-238. Because these particles are light, they can be carried many miles by wind currents. They are small enough to be inhaled into the terminal bronchi — the smallest air passages in the lungs, and can produce long-term health problems. External gamma radiation emitted from depleted uranium rounds can be as high as 200 millirads/hour, which is more than a year's dose from natural background radiation.

The Gulf War was the first time U.S. forces used depleted uranium rounds. M1A1, M1, and M60 tanks fired approximately 14,000 depleted-uranium antitank rounds. Air Force A-10 "tank-killer" planes fired about 940,000 30-millimeter depleted uranium rounds. Depleted uranium rounds were also used by the U.S. and its North Atlantic Treaty Organization (NATO) allies in Bosnia and Kosovo. Over 10,800 depleted uranium rounds were fired in Bosnia in 1994 and 1995, and about 31,000 depleted uranium rounds were fired in Operation Allied Force in Kosovo in 1999.[7] Since depleted uranium is very abundant, it is likely that terrorists will use rounds made of this material.

1.3.4 Other Potential Terrorist Uses for Radioactive Materials

While al-Qaida or another terrorist organization might gain access to a nuclear weapon of mass destruction, the likelihood is much greater that such organizations would first use more primitive weapons utilizing accessible radioactive materials including fuel from nuclear power plants, small radioactive sources used by the nuclear medicine industry, large radioactive sources used for medical radiography and teletherapy, radioactive devices used for food irradiation, radioactive components of instruments used to check welds in pipelines, and radioactive components used in down-hole geophysical survey instruments. These materials vary greatly in both radiological composition and activity levels — nuclear fuel has by far the highest activity levels of the group. The following are a few examples of potential ways terrorists could use these materials:

- Disposing of radioactive materials in a public drinking water supply
- Inserting radioactive materials into the heating and air conditioning ductwork of a large urban building to contaminate the air

- Dropping radioactive materials from a small aircraft into a densely populated area (city, outdoor sports event complex, outdoor concert hall)
- Parking a vehicle containing radioactive materials in a highly populated area
- Using radioactive materials in combination with conventional explosives to create a dirty bomb

It would be unlikely for any casualties to result from acute (short-term) effects of radiation exposure from such terrorist activities. The explosives used to distribute the radioactive material in dirty bombs would be much more likely to cause immediate or near-term casualties than the exposure to radiation. Any casualties from inhaling or ingesting small quantities of radioactive materials resulting from these types of terrorist attacks would likely take years to materialize. For this reason, such radioactive weapons are sometimes referred to as "weapons of mass disruption." Because only a small number of people are likely to be severely harmed or killed in the near-term by these attacks, the terrorist goals for these types of weapons are to frighten and disrupt the population. Such attacks could also cost governments millions of dollars to decontaminate surrounding buildings. If large-scale conventional explosives are used to detonate a dirty bomb, hundreds of casualties can easily result from the explosion alone. Recall that 168 people died in the 1995 bombing of the Alfred P. Murrah Federal Building in Oklahoma City.

The U.S. Nuclear Regulatory Commission oversees operations of more than 100 nuclear power reactors that generate electricity and 36 nonpower reactors located primarily at universities where they are used for research, testing, and training (Table 1.1). Nuclear fuel used to make dirty bombs could be obtained from any one of the U.S. nuclear power plants, or from plants scattered around the world. While the low-to-moderate radiation doses received from exposure to this type of attack would not be immediately life threatening, people near the detonation point could experience acute radiation exposure symptoms including headache, fatigue, weakness, anorexia, nausea, vomiting, diarrhea, lowered lymphocyte counts, decreases in white blood cell counts, and short-term cognitive impairment.

Long-term health effects from exposure to low-to-moderate doses of radiation include cancer of the thyroid, prostate, kidney, liver, salivary glands, and lungs; Hodgkin's disease; leukemia; and increased numbers of stillbirths and genetic defects. Concerns about potential long-term health effects often lead to anxiety and depression problems among those exposed to radiation.

The other types of radioactive materials cited in this section (medical industry and food industry sources) produce significantly lower activity levels than fuel from a nuclear power plant. However, these sources of radioactive materials may be appealing to terrorists because they are far more accessible. Thousands of hospitals, medical treatment facilities, and food industry plants scattered across the U.S. are protected by relatively low levels of security.

The following examples of terrorist events involving nuclear reactor fuel could produce large radiation doses (including emergency responders) resulting from the spread of contamination:

TABLE 1.1
U.S. Power and Nonpower Nuclear Reactors

Power Reactors

Arkansas Nuclear 1, Russellville, AR
Arkansas Nuclear 2, Russellville, AR
Beaver Valley 1, McCandless, PA
Beaver Valley 2, McCandless, PA
Braidwood 1, Joliet, IL
Braidwood 2, Joliet, IL
Browns Ferry 1, Decatur, AL
Browns Ferry 2, Decatur, AL
Browns Ferry 3, Decatur, AL
Brunswick 1, Southport, NC
Brunswick 2, Southport, NC
Byron 1, Rockford, IL
Byron 2, Rockford, IL
Callaway, Fulton, MO
Calvert Cliffs 1, Annapolis, MD
Calvert Cliffs 2, Annapolis, MD
Catawba 1, Rock Hill, SC
Catawba 2, Rock Hill, SC
Clinton, Clinton, IL
Columbia Generating, Richland, WA
Comanche Peak 1, Glen Rose, TX
Comanche Peak 2, Glen Rose, TX
Cooper, Nebraska City, NE
Crystal River 3, Crystal River, FL
D.C. Cook 1, Benton Harbor, MI
D.C. Cook 2, Benton Harbor, MI
Davis–Besse, Toledo, OH
Diablo Canyon 1, San Luis Obispo, CA
Diablo Canyon 2, San Luis Obispo, CA
Dresden 2, Morris, IL
Dresden 3, Morris, IL
Duane Arnold, Cedar Rapids, IA
Farley 1, Dothan, AL
Farley 2, Dothan, AL
Fermi 2, Toledo, MI
FitzPatrick, Oswego, NY
Fort Calhoun, Omaha, NE
Ginna, Rochester, NY
Grand Gulf 1, Vicksburg, MS
Harris 1, Raleigh, NC
Hatch 1, Baxley, GA
Hatch 2, Baxley, GA
Hope Creek 1, Wilmington, NJ
Indian Point 2, New York, NY
Millstone 2, New London, CT
Millstone 3, New London, CT
Monticello, Minneapolis, MN
Nine Mile Point 1, Oswego, NY
Nine Mile Point 2, Oswego, NY
North Anna 1, Richmond, VA
North Anna 2, Richmond, VA
Oconee 1, Greenville, SC
Oconee 2, Greenville, SC
Oconee 3, Greenville, SC
Oyster Creek, Toms River, NJ
Palisades, South Haven, MI
Palo Verde 1, Phoenix, AZ
Palo Verde 2, Phoenix, AZ
Palo Verde 3, Phoenix, AZ
Peach Bottom 2, Lancaster, PA
Peach Bottom 3, Lancaster, PA
Perry 1, Painesville, OH
Pilgrim 1, Plymouth, MA
Point Beach 1, Manitowoc, WI
Point Beach 2, Manitowoc, WI
Prairie Island 1, Minneapolis, MN
Prairie Island 2, Minneapolis, MN
Quad Cities 1, Moline, IL
Quad Cities 2, Moline, IL
River Bend 1, Baton Rouge, LA
Robinson 2, Florence, SC
Saint Lucie 1, Fort Pierce, FL
Saint Lucie 2, Fort Pierce, FL
Salem 1, Wilmington, DE
Salem 2, Wilmington, DE
San Onofre 2, San Clemente, CA
San Onofre 3, San Clemente, CA
Seabrook 1, Portsmouth, NH
Sequoyah 1, Chattanooga, TN
Sequoyah 2, Chattanooga, TN
South Texas 1, Bay City, TX
South Texas 2, Bay City, TX
Summer, Columbia, SC
Surry 1, Newport News, VA
Surry 2, Newport News, VA
Susquehanna 1, Berwick, PA
Susquehanna 2, Berwick, PA
Three Mile Island 1, Harrisburg, PA

-- continued

TABLE 1.1 (continued)
U.S. Power and Nonpower Nuclear Reactors

Power Reactors	
Indian Point 3, New York, NY	Turkey Point 3, Miami, FL
Kewaunee, Green Bay, WI	Turkey Point 4, Miami, FL
La Salle 1, Ottawa, IL	Vermont Yankee, Brattleboro, VT
La Salle 2, Ottawa, IL	Vogtle 1, Augusta, GA
Limerick 1, Philadelphia, PA	Vogtle 2, Augusta, GA
Limerick 2, Philadelphia, PA	Waterford 3, New Orleans, LA
McGuire 1, Charlotte, NC	Watts Bar 1, Spring City, TN
McGuire 2, Charlotte, NC	Wolf Creek 1, Burlington, KS
Nonpower Reactors	
Aerotest, San Ramon, CA	Rhode Island Atomic Energy Commission, Narragansett
Armed Forces Radiobiology Research Institute, Bethesda, MD	Texas A&M University, College Station
Cornell University, Ithaca, NY	U.S. Geological Survey, Denver, CO
	University of Arizona, Tuscon
Dow Chemical Company, Midland, MI	University of California, Irvine
General Electric Company, Pleasanton, CA	University of Florida, Gainesville
Idaho State University, Pocatello	University of Lowell, Lowell, MA
Kansas State University, Manhattan	University of Maryland, College Park
Massachusetts Institute of Technology, Cambridge	University of Michigan, Ann Arbor
	University of Missouri, Rolla
McClellan AFB, Sacramento, CA	University of Missouri, Columbia
National Institute of Standards & Technology, Gaithersburg, MD	University of New Mexico, Albuquerque
	University of Texas, Austin
North Carolina State University, Raleigh	University of Utah, Salt Lake City
Ohio State University, Columbus	University of Virginia, Charlottesville
Oregon State University, Corvallis	University of Wisconsin, Madison
Pennsylvania State University, University Park	Veterans Administration, Omaha, NE
Purdue University, West Lafayette, IN	Washington State University, Pullman
Reed College, Portland, OR	Worcester Polytechnic Institute, Worcester, MA
Rensselaer Polytechnic Institute, Troy, NY	U.S. Geological Survey, Denver, CO
	University of Arizona, Tuscon

- Setting off conventional explosives inside a nuclear power plant or spent nuclear fuel storage basin
- Crashing an airliner containing conventional explosives into a nuclear power plant or spent nuclear fuel storage basin
- Firing a small hand-held missile into an operating nuclear reactor
- Causing a nuclear reactor meltdown after taking over the control room

Experience from the 1986 Chernobyl reactor accident in the Ukraine shows the potential magnitude and impact of a terrorist attack on a nuclear power plant. The accident involved an explosion in a reactor that releases very high levels of radiation for miles surrounding the reactor site. Low levels of radiation were spread by wind currents throughout Europe and the rest of the world. According to Caldicott 2002, exposure to massive radiation levels immediately killed 32 plant workers and firefighters. Some references suggest that many more died in the following weeks and months due to high levels of radiation exposure, and large numbers of people suffer from Chernobyl-related illnesses. The world's scientists continue to be concerned about potential long-term genetic damage to future generations. Inhaling and ingesting radioactive particles and volatile isotopes such as iodine-131 from a nuclear accident are life threatening because the radiation comes in direct contact with living cells. The I-131 volatile radioactive isotope is particularly hazardous because it is absorbed quickly by the thyroid gland.

The consequences of a terrorist attack on a nuclear power plant may be determined by the parts of the reactor complex accessed by terrorists (auxiliary building, turbine building, control room), the specific nature of the damage resulting from the attack, and the remedial actions reactor operators can take. If the damage is limited to one function or component of the reactor, such as loss of coolant from the primary or secondary systems, the consequences may be minimized by remedial actions performed by the operators and built-in engineering safety features (emergency backup coolant system). However, if multiple functions are impacted by an attack (primary or secondary coolant systems and emergency backup coolant systems or cutoff of external electrical power and internal backup power), the consequences may be severe. The containment vessel of a reactor plays a critical role in limiting the consequences of an accident. The key defense against major radioactive emissions to the outside is keeping the containment intact as long as possible. Containment vessels are built with several safety devices (containment sprays). If a terrorist attack were to damage a containment vessel severely, especially in the early stages of the attack, the incident would likely release high levels of radiation emissions to the surrounding environment.

1.4 HISTORICAL DEVELOPMENT OF CHEMICAL WEAPONS

Chemical weapons were first used in modern warfare during World War I. On April 22, 1915, outside the Belgian village of Ypres, the German army released approximately 60 tons of chlorine gas from approximately 6000 pressurized gas cylinders into the winds, which carried clouds of the gas over the Allied forces. A second attack occurred 2 days later. Both attacks led to the choking deaths of approximately 10,000 troops. By the end of 1915, the Germans introduced a second gas agent, phosgene, which was ten times more toxic than chlorine.

Both phosgene and chlorine affect the pulmonary system and are thus called pulmonary agents. They damage the membranes in the lungs that separate the air sacs (alveoli) from the capillary blood vessels. As a result, plasma from the blood leaks into the air sacs. The sacs fill with fluid and this prevents air from entering. A person exposed to such an agent eventually suffocates after 2 to 24 hours. The victim initially experiences shortness of breath and severe coughing. Large amounts of yellowish (chlorine exposure) to clear (phosgene exposure) frothy sputum are produced along with significant irritation of the eyes and nose and burning of the throat.

By the time the Germans introduced phosgene, they were so proficient at producing and delivering chemical weapons that they became the world's leading manufacturers of such weapons. Great Britain quickly followed with the use of chemical weapons against the German army. Great skill was needed to deliver gas weapons because the poison could fall on the users' troops if wind direction changed or delivery was inappropriate. By the summer of 1917, the Germans introduced mustard gas, a chemical weapon classified as a blistering agent. As the name implies, the symptoms include blistering of the skin and other organs. The skin, eyes, lungs, gastrointestinal tract, mucous membranes, bone marrow, and other organs can be severely damaged.

These early weapons forced armies to defend themselves by using gas masks and protective suits, which severely impacted their ability to fight. By the end of World War I, more than 110,000 tons of chemical weapons were used by both sides. The number of injured measured over a million, and deaths numbered about 100,000.

1.4.1 Geneva Protocol

Because of the large number of deaths inflicted in World War I by chemical weapons, many nations wanted to ban their use. The League of Nations met at Geneva and developed a protocol signed by 38 nations in 1925 that eliminated use of chemical weapons for warfare. Despite the fact that the protocol had numerous loopholes and lacked provisions for the punishment of nations that violated the pact, no chemical weapons were used in World War II. However, new chemical weapons including tabun, sarin, soman, and others designated nerve agents were under development by the Germans in the 1930s and 1940s. Nerve agents are potent chemical weapons because they disrupt communications of nerves with the organs they stimulate. This interferes with basic bodily functions and can lead to death in 1 to 10 minutes after inhalation.

A nerve agent in liquid form is characteristically heavier than water; a nerve agent in vapor form is heavier than air and tends to sink toward the ground or the basement of a building. Although nerve agent vapors affect victims in a very short time, the range of effects varies greatly, depending on degree of exposure. Exposure to nerve agents initially affects airways and parts of the face that come into contact with the agent: the eyes, nose, and mouth. The pupils become small pinpoints, the eyes become red, and vision is blurred. Some victims also experience eye pain, headache, nausea, and vomiting. Rhinorrhea (discharge of nasal mucus) and excessive salivation are also common symptoms of exposure.

If a nerve agent is inhaled, airways can become constricted, inducing coughing fits or shortness of breath. If a sufficient quantity of agent is inhaled, sudden loss of consciousness and convulsions may ensue. A victim may stop breathing within a few minutes. A very small amount of a nerve agent like sarin can produce a phenomenon called fasciculation (muscular twitching). With greater exposure, involuntary defecation and urination may also occur.

1.4.2 A New Age of Chemical Weapons Development

After the defeat of the Germans in 1945, the Allied forces discovered chemical weapon manufacturing plants and large stores of chemical weapons in several German cities. They seized the weapons and began programs for developing their own chemical weapons. The Russians moved entire chemical plants from German cities to Volgograd, where they began research and development on new chemical weapons. The end of World War II brought about new chemical, biological, and nuclear arms races.

Many new nerve agent weapons were developed from ordinary insecticides and pesticides. Common insecticides such as malathion, sevin, and many others are actually nerve agents. In the early 1950s, Great Britain discovered a new nerve agent that was magnitudes more lethal than any other known substance. Its code name was VX. It was more lethal and more persistent (it remains a liquid for more than 24 hours) than other nerve agents. It can enter the body either by inhalation or skin contact. Entry through the skin was possible because VX is nonvolatile and persistent. The U.S. cooperated with Great Britain and eventually took over large-scale production. VX production and stockpiling continued into the 1960s.

One of the better-known U.S. chemical agents used during the Vietnam War was Agent Orange, a herbicide used to defoliate vegetation. Agent Orange contains varying quantities of dioxin (tetrachlorodibenzo-dioxin). It was determined to be so dangerous that the U.S. Environmental Protection Agency prohibited its use in 1986. Animal studies linked it to non-Hodgkin's lymphomas (cancers of the lymph system), sarcomas (soft tissue cancers), carcinomas (epithelioid cancers), and a host of other diseases.

1.4.3 Prohibition of Development, Production, and Stockpiling of Biologic and Toxic Weapons

In 1972, more than 100 countries including the U.S. signed the Convention on the Prohibition of the Development, Production, and Stockpiling of Biologic and Toxic Weapons and Their Destruction — a measure designed to limit further development or use of biological and chemical weapons. Unfortunately, the accord has been breached several times.

In 1988, Libya built a chemical weapons plant and disguised it as a pharmaceutical factory. Iraq used mustard gas in its long war with Iran and used mustard gas and toxic nerve agents against its own dissident Kurdish population. Saddam Hussein's stockpile of weapons of mass destruction is thought to include many other chemical and biological warfare agents. Part of the problem is that companies in the West are selling so-called dual-use technologies to countries like Iraq and Libya, where a plant for making pesticides can readily be converted to produce chemical weapons. Other countries that used chemical warfare agents in the 1980s were Afghanistan, Cambodia, Iran, and Laos.

In March 1995, a Japanese religious cult, Aum Shinriyko, released sarin nerve gas in the Tokyo subway system. Thousands were injured and 11 people were killed.

This use of chemical weapons on a civilian population underscores the relative availability and ease of use of such weapons. It also raises both public and governmental consciousness.

1.4.4 Weaponization and Delivery of Chemical Agents

All chemical agents must be weaponized prior to delivery. The process involves numerous steps. First, an agent must be made in sufficient quantity to match the size of the desired target. It must be stored temporarily and stabilized to prevent evaporation and degradation. Addition of thickeners to increase the viscosity of liquid agents and a carrier agent to improve dispersion of the chemical is usually required. The chemical agent must then be inserted into appropriate munitions or dissemination devices to be used on a target. These devices can be categorized as explosive, pneumatic, or mechanical. Whatever the dissemination device, the usual goal is to aerosolize the chemical agent to a particulate size of 1 to 7 microns by using sophisticated delivery systems such as munitions devices usually only available to governments or unsophisticated devices like aerosol generators such as deodorant containers or garden sprayers. Other factors that affect the weaponization and delivery of a chemical weapon include:

- Air and ground temperature: higher temperatures generally cause faster evaporation.
- Humidity: high humidity may cause enlargement of particle size, thus reducing effectiveness.
- Precipitation: heavy rain dilutes and disperses chemical weapons; snow increases persistence.
- Wind speed: wind can disperse vapors, aerosols, and liquids, thus affecting the target.
- Building construction: buildings can absorb or adsorb agents and also offer protection.
- Nature of terrain: woodlands and hills create greater turbulence of low-lying clouds.

1.5 HISTORICAL DEVELOPMENT OF BIOLOGICAL WEAPONS

The first recorded use of biological warfare occurred in ancient Assyria in the Sixth Century B.C. The Assyrians poisoned the wells of their enemies with a rye fungus known as ergot. For many years, this constituted biological warfare. More extensive biological warfare began in 1346 A.D., when Tatars besieging Italian traders in a citadel in Crimea threw corpses of plague victims over the wall. Evidence indicates Russians used the same biological warfare tactic of hurling corpses of plague victims over the city walls of Reval in a battle with the Swedes in 1710.

In 1763, an American captain presented two smallpox-laden blankets and a handkerchief to Native Americans as a gesture of friendship. An English general also gave blankets contaminated with smallpox to Native Americans loyal to the

French. Since then, developments in biological weapons have kept in step with new developments in medical science.

1.5.1 Modern Biological Weapons

During World War I, the Germans were suspected of using germ warfare to spread cholera in Italy, plague in Russia, and other agents in Great Britain. During World War II, all the major powers conducted biological warfare research. Germany and Japan actively used biological weapons. A Nazi doctor named Josef Mengele is known to have injected prisoners at Auschwitz with typhoid and tuberculosis germs. In 1936, the Japanese formed a large biological warfare unit called the Epidemic Prevention and Water Supply Unit, or Unit 731, that conducted experiments on captured Chinese, Soviet, and American soldiers and may have used biological weapons on Chinese cities in the 1930s and 1940s.

While the U.S. has had biological weapons, it has not used them. The U.S. Army developed Camp Detrick, Maryland, into a site for biological research and development. The site manufactured anthrax and botulinum bombs in the event they were needed during World War II. After the war, knowing that biological weapons could become very important, the U.S. shielded Unit 731 from prosecution for war crimes in exchange for its research data and expertise.

After World War II, a biological weapons arms race proceeded in parallel with the nuclear weapons arms race. Biological weapons became far more common at a much faster rate than nuclear weapons because of their low cost and minimal technical demands. In 1950, the U.S. biological weapons program was in high gear to prepare for the possible use of biological weapons during the Korean War. The U.S. Army Medical Research Institute of Infectious Diseases was created in 1969 — the same year in which President Nixon renounced biological weapons and limited future research to defensive measures.

Throughout recent decades, efforts were made to limit or ban biological weapons, but several nations ignored the 1972 Convention on the Prohibition of the Development, Production, and Stockpiling of Biologic and Toxic Weapons and their Destruction. Between 1975 and 1983, Laos and Cambodia came under attack by planes and helicopters that delivered "Yellow Rain" suspected to contain lethal T-2 mycotoxins.

In 1991, the Iraqi government announced that it conducted biological weapons research. Unfortunately, the U.S. did not destroy all its facilities during the Persian Gulf War. In 1995, Iraqi General Hassan, who defected to the United States, announced that the Iraqi biological weapons unit was far more advanced than previously believed. It is now believed that Iraq, Iran, Libya, Syria, North Korea, Taiwan, Israel, Egypt, Vietnam, Russia, Cuba, Laos, South Korea, India, Bulgaria, South Africa, and China have manufactured biological weapons.

1.5.2 Current Genetic and Molecular Engineering

The pace of biological weapons development may well be accelerating, due to innovations in genetic and molecular engineering. A team of Australian scientists announced in January 2001 that they created a killer mouse pox virus unintentionally. Working with a weakened virus that usually only makes mice slightly ill, the researchers wanted to create a contraceptive that would trigger an immune response to a mouse's own eggs. They genetically altered one part of the mouse pox virus by inserting a gene that directs the production of interleukin-4, an important immune system chemical in humans and mice. Instead of acting as a contraceptive, the virus destroyed the immune systems of the mice. The mouse pox virus cannot harm humans, but the implication of the research is that any virus that afflicts human beings, such as a common cold virus, could be turned into a killer virus in the same way.

It is possible that the smallpox virus will be made even deadlier than it was in its natural form. Smallpox, which is estimated to have killed half a billion people throughout history, has been eradicated in nature. The last recorded case occurred in Africa in 1977, and most countries stopped vaccinating against it in 1980. As a result of medicine's victory, practically all stocks of smallpox viruses and vaccines were supposed to have been destroyed by April 1999. Shortly before that, President Clinton ordered that a collection of smallpox viruses and vaccines be maintained for security purposes because U.S. intelligence agencies found that collections were maintained by Russia, Iraq, Iran, North Korea, and other countries. If another country or terrorist group unleashed smallpox in the U.S. in a genetically altered form, we would desperately need a collection of virus strains and vaccines to serve as a basis to find ways to limit the damage. In June 2000, groups of 6- to 12-year old children in Vladivostok, Russia, found and played with negligently discarded ampoules of weakened smallpox. Because the smallpox was in a weakened form from which vaccines could have been made, the children survived infection, but their faces were permanently scarred. Initially the doctors who treated the children did not know what to do. Since smallpox was eradicated in nature more than 20 years earlier, they did not immediately recognize the symptoms.

In January 2001, at least 25 similar threats were received across Canada and the U.S. A typical case occurred at a Wal-Mart in Victoria, British Columbia, when a letter claiming to contain anthrax was received. Fortunately the letter did not contain anthrax, but before that could be conclusively established, the clerk who opened the letter received a precautionary dose of ciprofloxacin.

1.5.3 Post-September 11 Anthrax Attacks

Until October 2001, no deaths from bioterrorism were reported in the U.S. On September 11, 2001, after the attacks on New York City and Washington, the Centers for Disease Control recommended that the nation increase its surveillance for unusual disease occurrences or clusters, asserting that they could be sentinel indicators of bioterrorist attacks. As predicted, cases of anthrax were reported in Florida, New York City, the District of Columbia, and New Jersey. Over the years, it has become evident that not only is biological warfare attractive to governments, it is equally attractive to terrorist cells because the agents are relatively inexpensive and easy to make.

Hundreds of biological agents could serve as terrorist weapons. Some are more likely candidates than others. The U.S. Department of Defense published a list of 17 of the most likely biological weapons divided into four categories: (1) bacteria such as anthrax and plague germs; (2) viruses such as smallpox and encephalitis; (3) hemorrhagic fevers such as Ebola, Lassa and Rift Valley fevers; and (4) toxins such as botulinum, fungal toxins, and ricin that attack the central nervous system. See Sections 2.2 and 2.3 in the next chapter for specific details on these biological weapons.

1.5.4 Agroterrorism

Instead of trying to spread disease directly in a population, terrorists might attack indirectly by poisoning a country's food supplies. Agroterrorism, as it is coming to be called, is emerging as an increasing threat. U.S. Department of Agriculture administrator Floyd Horn said "a biological attack (on America's crops or livestock) is quite plausible."

Many diseases such as foot-and-mouth disease that take a devastating toll on food animals do not affect humans. Spreading such diseases is technically easy and a low-risk strategy for terrorists, who would need to take no special precautions in handling and deploying disease-causing germs. However, the costs are enormous to a victim country. The 2001 outbreak of foot-and-mouth disease in Great Britain cost several billion dollars and profoundly demoralized the entire country. The epidemic started on a single farm. It spread with devastating speed because the virus that causes the disease can be transmitted easily, including by airborne particles and even on shoes that track dirt from infected areas to noninfected areas. Although this outbreak was natural and was not linked to terrorism, it suggested the scope of damage that terrorists could inflict on an entire country or large region simply by infecting herds on a few farms.

To bring attention to a cause or extract money, terrorists could threaten a disease outbreak as a type of biological protection racket. Alternatively, organized criminals might use agroterror to manipulate commodities markets. Advance knowledge that certain crops or herds will be tainted could serve as a means to ensure huge profits from trading in future contracts on untainted crops and herds. It is important to note that the spread of foot-and-mouth disease in Great Britain in 2001 was helped along by government cuts in inspection and food safety measures. Great Britain slashed budgets for surveillance of its food and livestock. In 1991, it had 43 regional animal health offices and 330 government veterinarians. Today it has half as many offices, fewer veterinarians, and fewer government laboratories for disease testing. The virus causing foot-and-mouth disease was the first virus discovered by science in 1898. More than a century later, we can do little more than quarantine and destroy as many as hundreds of thousands of infected animals to halt an epidemic of this terrible disease.

1.6 CONVENTIONAL EXPLOSIVES AVAILABLE FOR DISPERSING AGENTS

Terrorists will likely consider the option of using conventional explosives to distribute nuclear, chemical, or biological agents. For this reason, the following section presents details on common commercially available explosives and many used by the military. This information has been provided to assist emergency responders in recognizing various types of explosives, highlight explosives with high degrees of sensitivity (they are easily detonated when exposed to shock, friction, electric spark), and introduce analytical methods that can be used to test for the presence of explosives. Much of the information in this section was derived from one or more of the following references:

- Akhavan, J., *The Chemistry of Explosives*, Royal Society of Chemistry, Letchworth, U.K., 1998.
- Suceska, M., *Test Methods for Explosives,* Springer-Verlag, New York, 1995.
- Yinon, J. and Zitrin, S., *Modern Methods and Applications in Analysis of Explosives*, John Wiley & Sons, New York, 1996.
- Pickett, M., *Explosives Identification Guide*, Delmar Publishers, New York, 1999.

The reader is encouraged to refer to these volumes for additional details not provided in this section.

The four general classifications of explosives are (1) primary explosives, (2) secondary explosives, (3) propellants, and (4) pyrotechnics. Primary explosives are readily ignited or detonated when subjected to heat or shock. They are generally used to assist detonation of less sensitive but more powerful secondary explosives by means of high-velocity shock waves. Propellants are initiated by a flame, spark, shock, friction, or high temperature. They only deflagrate (burn with a flame or spark) and do not detonate. Pyrotechnics (flares and fireworks) are often initiated by a primer and emit colored smoke, noise, and bright colored light.[8]

Explosives are generally detonated with the assistance of initiating devices (initiators) that are activated by external stimulation sources (friction, spark, or flame), and may include (1) primers, (2) detonators, (3) electric detonators, (4) safety fuses, and (5) detonating cord.

Primers are used for the ignition of propellants and pyrotechnics. They are categorized as percussion primers, used mainly for the ignition of propellants, and stab primers, used in explosive trains of fuses and electric primers.

Detonators are used to detonate high explosives. Stab detonators are initiated by sharp firing pins and are used in explosive trains of different types of fuses. Flash detonators are initiated by flames produced by safety fuses, primers, or delay elements. A special type of flash detonator ignited by the flame of a safety fuse is called a blasting cap. Detonators are primarily composed of three types of explosives including sinoxid mixtures, lead azide-based mixtures, and mercury fulminate-based mixtures.

Electric detonators are also used for detonation of high explosive charges. They are similar in design to other types of detonators except for the presence of an electric fusehead consisting of a bridgewire made of chromium and nickel. The bridgewire is covered by a heat-sensitive pyrotechnic mixture protected by varnish insulation. Standard fuseheads have electrical resistance of 1.2 to 1.4 ohms and are activated by a current of 0.7 amps. Electric detonators can have instantaneous or delay actions.

Safety fuses are used for the direct ignition of propellants, pyrotechnics, and primary explosives. A safety fuse is composed of a black powder core, three layers of cotton or jute yarn wound around the core, bitumen impregnation, and plastic coating. The burn rate for a safety fuse is usually around 120 ± 10 seconds/meter.

A detonating cord is used for simultaneous detonation of a number of explosive charges. The cord is comprised of a core (often made of Pentrit), cotton threads around a core, and a plastic coating. A detonating cord with a Pentrit core has a detonation velocity of approximately 6500 meters/second and is initiated by a blasting cap.[9]

1.6.1 Primary Explosives

Primary explosives differ from secondary explosives in that they undergo a rapid transition from burning to detonation and have the ability to transmit the detonation to less sensitive (but more powerful) secondary explosives. Primary explosives have high degrees of sensitivity to initiation through shock, friction, electric spark, or high temperature. and explode whether confined or unconfined. Some widely used primary explosives include lead azide, silver azide, tetrazene, lead styphnate, mercury fulminate, and diazodinitrophenol. Nuclear weapon applications normally limit the use of primary explosives to lead azide and lead styphnate.

1.6.1.1 Lead Azide

Lead azide (PbN_6) is a colorless to white crystalline explosive. It is widely used in detonators because of its high capacity for initiating secondary explosives to detonation. However, since lead azide is not particularly susceptible to initiation by impact, it is not used alone in initiator components. It is used in combination with lead styphnate and aluminum for military detonators, and is used often in a mixture with tetrazene. It is compatible with most explosives and priming mixture ingredients. Contact with copper must be avoided because it leads to formation of extremely sensitive copper azide.

This primary explosive is created by adding lead acetate to a solution of sodium or ammonium azide. Lead azide has a good shelf life in dry conditions but is unstable in the presence of moisture, oxidizing agents, and ammonia. It is less sensitive to impact than mercury fulminate, but more sensitive to friction. Since lead azide is a nonconductor, it may be mixed with flaked graphite to form a conductive mixture for use in low-energy electronic detonators.

1.6.1.2 Silver Azide

Silver azide (AgN_3) is a white-colored crystalline explosive. It requires less energy for initiation than lead azide, and fires with a shorter time delay. This primary explosive decomposes under the influence of ultraviolet radiation. If the ultraviolet radiation is too intense, silver azide can explode. It is created by the action of sodium azide on silver nitrate in an aqueous solution.

1.6.1.3 Tetrazene

Tetrazene ($C_2H_8N_{10}O$) is a pale yellow crystalline explosive generally used in ignition caps, where a small amount is added to the explosive composition to improve its sensitivity to percussion and friction. Tetrazene is not suitable for filling detonators because its compaction properties make the transition from burning to detonation very difficult. This primary explosive is stable in ambient temperatures. Its ignition temperature is lower and it is slightly more sensitive to impact than mercury fulminate.

1.6.1.4 Lead Styphnate

Lead styphnate ($C_6H_3N_3O_9Pb$) is a weak primary explosive because of its high metal content and therefore is not used alone in the filling of detonators. It is used in ignition caps or can be mixed with lead azide and aluminum for use in detonators. It can also be mixed with graphite to enhance its electrical conductivity. Lead styphnate is usually prepared by adding a solution of lead nitrate to magnesium styphnate. It is practically insoluble in water and most organic solvents. It is very stable at room and elevated (75°C) temperatures. It is exceptionally resistant to nuclear radiation and can easily be ignited by a flame or electric spark.

1.6.1.5 Mercury Fulminate

Mercury fulminate ($C_2N_2O_2Hg$) is one of the most important primary explosives. It is usually found in the form of a gray powder, is sensitive to impact and friction, and is easily detonated by sparks and flames. It is desensitized by the addition of water, but is very sensitive to sunlight. It reacts with metals in moist environments. It is created by treating a solution of mercuric nitrate with alcohol in nitric acid. Its most important explosive property is that it easily detonates after initiation.[10]

1.6.1.6 Diazodinitrophenol

Diazodinitrophenol ($C_6H_2N_4O_5$) has been used as an initiator in industrial blasting caps. It is less sensitive to impact, friction, and electrostatic energy than other primary explosives. It is orange-yellow in color and is sparingly soluble in water.[11]

1.6.2 Secondary Explosives

Secondary explosives (also known as high explosives) are different from primary explosives in that they cannot be detonated readily by heat or shock and are generally more powerful. Secondary explosives can be initiated to detonation only by a shock produced by the explosion of a primary explosive. Widely used secondary explosives include trinitrotoluene (TNT), tetryl, picric acid, nitrocellulose, nitroglycerine, nitroguanidine, cyclotrimethylenetrinitramine (RDX), cyclotetramethylenetetranitramine (Octogen or HMX), 1,3,5-triamino-2,4,6-trinitrobenzene (TATB), pentaerythritol tetranitrate (PETN), and hexanitrostilbene (HNS)

1.6.2.1 Trinitrotoluene (TNT)

Trinitrotoluene or TNT ($C_7H_5N_3O_6$) is a pale yellow crystalline solid widely used in military explosives before World War I and still used. TNT has low sensitivity to impact and friction, good chemical and thermal stability, good compatibility with other explosives, a low melting point for casting, and fairly high explosive power. TNT is the most important explosive for blasting charges. It is widely used for commercial explosives and is much safer to produce and handle than nitroglycerine and picric acid. It can be loaded into shells by casting and pressing. It can be used alone or by mixing with other components such as ammonium nitrate to yield amatol, with aluminum powder to yield tritonal, or with RDX to yield cyclonite. One of its primary disadvantages is the leaching out of the isomers of dinitrotoluenes and trinitrotoluenes. This often occurs in the storage of projectiles containing TNT, particularly in the summer. This leaching can cause the formation of cracks and cavities in the shells and lead to premature detonation. TNT is almost insoluble in water, sparingly soluble in alcohol, and dissolves in benzene, toluene, and acetone. It darkens in sunlight and is unstable in alkalis and amines.

1.6.2.2 Tetryl

Tetryl ($C_7H_5N_5O_8$) is a pale yellow crystalline solid. It is moderately sensitive to initiation by friction and percussion and is used in the form of pressed pellets as primers for explosive compositions that are less sensitive to initiation. It is slightly more sensitive than picric acid and considerably more sensitive than TNT. In the early 1900s, tetryl was used as base charges for blasting caps but now has been replaced by PETN and RDX. During World War II, tetryl was used as a component of explosive mixtures.

1.6.2.3 Picric Acid

Picric acid ($C_6H_3N_3O_7$) is a yellow crystalline solid prepared by reacting phenol with nitric acid. Its explosive power (strength and velocity of detonation) is somewhat superior to that of TNT. Picric acid was historically used in grenade and mine fillings but had a tendency to form impact-sensitive metal salts (picrates) with the metal walls of shells. Filling of mines and grenades was a hazardous process because relatively high temperatures were needed to melt the picric acid.

1.6.2.4 Nitrocellulose

Nitrocellulose [$C_6H_7O_2(OH)_x(ONO_2)_y$ where $x + y = 3$] materials prepared from cotton are fluffy white solids that ignite around 180°C and do not melt. This explosive dissolves in organic solvents to form a gel that has good physical properties and is an essential ingredient of gun propellants, double-base rocket propellants, and gelatin and semigelatin commercial blasting explosives. Nitrocellulose is sensitive to initiation by percussion or electrostatic discharge and can be desensitized by the addition of water. Its thermal stability decreases with increasing nitrogen content. Its chemical stability depends on the removal of all traces of acid in the manufacturing process.

1.6.2.5 Nitroglycerine

Nitroglycerine ($C_3H_5N_3O_9$) is a very powerful secondary explosive with a high shattering effect. It is one of the most important and frequently used components of gelatinous commercial explosives. Nitroglycerine also provides a source of high energy in propellant compositions, and in combination with nitrocellulose and stabilizers, it is the principal component of explosive powders and solid rocket propellants. It is yellow in color and is created by adding glycerol to a mixture of sulfuric and nitric acids.

1.6.2.6 Nitroguanidine

Nitroguanidine ($CH_4N_4O_2$) is a white fiber-like crystalline solid, also known as picrite. It has a high velocity of detonation, low heat and temperature of explosion, and high density. It can serve as a secondary explosive but is also suitable for use in flashless propellants because of its low heat and explosion temperature. Flashless propellants contain mixtures of nitroguanidine, nitrocellulose, nitroglycerine, and nitrodiethyleneglycol that form colloidal gels. Nitroguanidine is relatively stable below its melting point (257°C), but decomposes immediately upon melting to form ammonia, water vapor, and solid products. It is soluble in hot water and alkalis.

1.6.2.7 Cyclotrimethylenetrinitramine (RDX)

Cyclotrimethylenetrinitramine (also known as Hexogen, Cyclonite, and RDX) is a white crystalline solid with a melting point of 205°C. RDX is more chemically and thermally stable than PETN. Pure RDX is very sensitive to initiation by impact and friction. It is desensitized by coating its crystals with wax, oils, or grease. It can also be compounded with mineral jelly and similar materials to produce plastic explosives. Insensitive explosive compositions containing RDX can be achieved by embedding the RDX crystals in a polymeric matrix. This composition is referred to as a polymer-bonded explosive (PBX). RDX has high chemical stability and great explosive power compared with TNT and picric acid. It is difficult to dissolve RDX in organic liquids, but it can be re-crystallized from acetone. Its high melting point makes it difficult to use in casting. However, when mixed with TNT, a pourable mixture can be obtained.

1.6.2.8 Cyclotetramethylenetetranitramine (Octogen or HMX)

Cyclotetramethylenetetranitramine or HMX ($C_4H_8N_8O_8$) is a white crystalline solid with a melting point of 285°C. HMX is superior to RDX in that its ignition temperature is higher and its chemical stability is greater. However, its explosive power is somewhat lower than that of RDX. HMX is insoluble in water and behaves similarly to RDX with respect to chemical reactivity and solubility in organic liquids. However, HMX is more resistant to attack by sodium hydroxide and is more soluble in 55% nitric acid and 2-nitropropane than RDX.

1.6.2.9 1,3,5-triamino-2,4,6-trinitrobenzene (TATB)

TATB or 1,3,5-triamino-2,4,6-trinitrobenzene ($C_6H_6N_6O_6$) is a yellow-brown crystalline solid that has excellent thermal stability and is known as a heat-resistant explosive. TATB has a decomposition point of 325°C. Its molecular arrangement provides lubricating and elastic properties.

1.6.2.10 Pentaerythritol Tetranitrate (PETN)

Pentaerythritol tetranitrate (PETN) is a colorless crystalline solid that is very sensitive to initiation by a primary explosive. It is a powerful secondary explosive that has a great shattering effect. It is used in commercial blasting caps, detonation cords, and boosters. PETN is not used in its pure form because it is too sensitive to friction and impact. It is usually mixed with plasticized nitrocellulose or with synthetic rubbers to form PBXs. The most common form of explosive composition containing PETN is Pentolite, a mixture of 20 to 50% PETN and TNT. PETN can be incorporated into gelatinous industrial explosives. The military has in most cases replaced PETN with RDX because RDX is more thermally stable and has a longer shelf life. PETN is insoluble in water, sparingly soluble in alcohol, ether, and benzene, and soluble in acetone and methyl acetate.

1.6.2.11 Hexanitrostilbene (HNS)

Hexanitrostilbene or HNS ($C_{14}H_6N_6O_{12}$) is a heat-resistant yellow crystalline solid explosive. HNS is also resistant to radiation, insensitive to electric sparks, and less sensitive to impact than tetryl. It is used in heat-resistant booster explosives and has been used in stage separations in space rockets and for seismic experiments on the moon. Its melting temperature is 316 °C.[12]

1.6.3 Propellants

Propellants are explosives that undergo rapid and predictable combustion without detonation. They release large volumes of hot gas that can be used to propel a projectile such as a bullet or missile. To produce gas quickly, a propellant must carry its own oxygen along with suitable quantities of fuel (carbon, hydrogen). Gun propellants are typically homogeneous — the oxygen and fuel are present in the same molecule. Missile propellants are typically heterogeneous in that the oxygen and fuel are separate compounds.

1.6.3.1 Gun Propellants

Gun propellants have traditionally been fabricated from nitrocellulose-based materials. These fibrous materials can be manufactured in granular or stick forms (grains) to provide a constant burning surface without detonation. The size of the propellant grains is based on the size of the gun. Larger guns require larger grains that provide longer burn times. Single-based propellants are used in all types of guns ranging from pistols to artillery weapons. The propellants consist of approximately 90% nitrocellulose (nitrogen content of 12.5 to 13.2%).

Double-based propellants are a mixture of nitrocellulose and nitroglycerine. This mixture increases the pressure of the gas inside the gun barrel. Double-base propellants are used in pistols and mortars. Some disadvantages of using double-base propellants is the erosion that this mixture causes to the gun barrel due to higher burn temperatures and the presence of muzzle flash.

Triple-based propellants are mixtures of nitroguanidine, nitrocellulose, and nitroglycerine. The mixture reduces the muzzle flash observed with double-based propellants, reduces the burn temperature, which protects the gun barrel, and increases the gas volume. Triple-based propellants are used in tank guns, large caliber guns, and some naval weapons.

1.6.3.2 Rocket Propellants

Rocket propellants are made in solid and liquid forms. Solid rocket propellants are similar to gun propellants in that they are formed into grains. For short-range missiles, the grains are larger and fewer in number than in gun cartridges, and they are designed to burn over their entire surface to yield a high mass burn rate. Long-range missiles usually only contain one or two large grains. Liquid rocket propellants use fuels that can burn in the absence of external oxygen (hydrazine) or must contain both a fuel (methanol, kerosene) and an oxidizer (nitric acid).

1.6.4 Pyrotechnics

Pyrotechnics are often used in the production of fireworks and primers. A pyrotechnic contains both a fuel and an oxidizer formulated to produce a lot of energy. The energy is then used to produce a flame or glow (matchstick) or combined with other volatile substances to produce smoke and light (fireworks) or large quantities of gas (firework rockets).

Pyrotechnic primers are often used to ignite gun propellants. A flame is generated from the primer when it is struck with a metal firing pin, which in turn ignites the gun propellant. Pyrotechnic primers typically contain potassium chlorate, lead peroxide, antimony sulfide, or trinitrotoluene. Pyrotechnics such as black powder, tetranitrocarbazole combined with potassium nitrate, and others are often used to provide time delays for explosives. Delay compositions that have fast burn rates are often used in projectiles and bombs that explode on contact. Those with slow burn rates are used in ground chemical munitions such as tear gas and smoke grenades. Pyrotechnic compositions such as zinc dust mixed with hexachloroethane and aluminum are used to produce smoke for use in flares, screening and camouflaging, special effects for theaters and films, and military training aids.

1.6.5 Other Compounds Used in Explosives

Aluminum powder is frequently added to explosive and propellant compositions to improve their efficiency. Ammonium nitrate is the most important raw material used in the manufacture of commercial explosives and it is also used as an oxygen source in rocket propellant compositions. Commercial blasting explosives contain ammonium nitrate, wood meal, oil, and TNT. Small glass or plastic spheres containing oxygen can be added to emulsion slurries to increase sensitivity to detonation. Phlegmatizers (waxes) can be added to explosives to aid processing and reduce impact and friction sensitivity of highly sensitive explosives.

1.6.6 Initiation Techniques

An explosive device is initiated or detonated by an explosive train — an arrangement of explosive components by which the initial force from the primary explosive is transmitted and intensified until it reaches and sets off the main explosive composition. Most explosive trains contain a primary explosive as the first component. The second component in the train will depend on the type of initiation process required for the main explosive composition. If the main explosive composition is to be detonated, the second component of the train will burn to detonation so that it imparts a shockwave to the main composition. This type of explosive train is known as a detonator. Detonators can be initiated by electrical means, friction, flash, or percussion.

If an explosive train is only required to ignite a main composition, an igniter that produces a flash instead of a detonation is used. Explosives of this kind are known as deflagrating explosives. Similar to detonators, igniters can be initiated by electrical means, friction, flash, or percussion. An example of an igniter is a squib, a small explosive device loaded with an explosive that deflagrates. Its output is primarily heat.[8]

1.6.7 Analytical Methods for Testing for Explosives

Analytical procedures used to identify specific explosive types may be performed before or after detonation. Running analytical tests on explosives prior to detonation is generally performed for the purpose of verifying that the material in question is an explosive and learning how to properly handle and dispose of the material. Running analytical methods following detonation are often performed with the objective of trying to identify who was responsible for the detonation. In that case, hand swab samples may be collected from the scene of the explosion and from suspects, their clothing, vehicles, houses, etc. This information could later be used as courtroom evidence if it provides a match. The type of explosive used may ultimately indicate whether the explosion was a criminal or terrorist act.

Some of the more effective chromatographic methods for testing for explosives, and their advantages and disadvantages are summarized in Table 1.2. When attempting to identify hidden explosives, a number of vapor detection methods should be considered including the use of (1) trained dogs that have highly sensitive olfactory systems (effective for TNT, EGDN, nitroglycerine, plastic explosives, smokeless

TABLE 1.2
Analytical Methods for Testing for Explosives

Analytical Method	Advantages	Disadvantages
Thin layer chromatography (TLC)	Acceptable screening method for organic explosives, particularly for postexplosion residue analysis Simple, rapid, inexpensive, and relatively sensitive Can be used to confirm presence of explosives in hand swabs Effective in testing for ethylene glycol dinitrate (EGDN), nitroglycerine, TNT, PETN, RDX, tetryl, HMX, nitrocellulose, and others	TLC is a screening method; a small percentage (5 to 10%) of the samples should be confirmed by a more sophisticated analytical method such as gas chromatography or high performance liquid chromatography
Gas chromatography (GC)	Well established organic chemistry technique While a flame ionization detector (FID) can be used with this method to detect explosives with large numbers of carbon atoms such as nitroaromatic compounds, GC generally is not recommended for detection of other types of explosives Electron capture detector (ECD) and thermal energy analyzer (TEA) detector are preferred over FID ECD is a very sensitive detector for explosives containing nitrogen atoms; ECD is effective in testing for explosives in a matrix such as postexplosive debris; ECD can detect industrial and military explosives (e.g., EGDN, nitroglycerin, 2,4-dinitrotoluene (DNT), TNT, PETN, RDX, tetryl) in the low nanogram range	Some explosives are not volatile enough to be analyzed via GC; the relatively high temperatures required can cause decomposition of some explosives (e.g., nitrate esters, nitramines); excessive contamination often present in hand swabs from postexplosion debris can interfere with optimum performance of some detectors

-- continued

TABLE 1.2 (CONTINUED)
Analytical Methods for Testing for Explosives

Analytical Method	Advantages	Disadvantages
High performance liquid chromatography (HPLC)	UV detectors, ECD, and TEA have been used with HPLC to analyze explosives The UV detector was the first to be used with HPLC and is still the most popular; some explosives can be detected in the nanogram range with the UV detector, including TNT, RDX, tetryl, PETN, tetrytol, nitroglycerine, EGDN, and HMX Some explosives that can be detected in the nanogram range using ECD include nitrobenzene, 2,4-DNT, 2,6-DNT, EGDN, PETN, and nitroglycerine TEA is becoming a more popular detector; it can identify a large variety of explosives including TNT, DNT, RDX, HMX, EGDN, PETN, and nitroglycerine	Restrictions caused by low volatility and thermal instability of some explosives
Supercritical fluid chromatography (SFC)	Overcomes some difficulties of GC and HPLC such as restrictions caused by low volatility and thermal instability of many organic compounds (GC) and long analysis times (HPLC) Permits high resolution at low temperatures with short analysis times With a TEA detector, SFC can detect a number of explosives including nitroglycerine, EGDN, PETN, mannitol hexanitrate, RDX, HMX, TNT, tetryl, hexanitrobibenzyl, trinitro-*m*-xylene, 2-nitrotoluene, picric acid, and 2,4-DNT	Not as common as GC or HPLC

Source: Yinon, J. and Zitrin, S., *Modern Methods and Applications in Analysis of Explosives*, John Wiley & Sons, New York, 1996. With permission.

powder, black powder), and (2) a portable gas chromatograph with ECD; a portable ion mobility spectrometer. Other analytical methods that should also be considered for testing for explosives include mass spectrometry, nuclear magnetic resonance spectrometry, ion chromatography, and scanning electron microscopy/energy-dispersive x-ray spectrometry. For additional details on these methods, refer to Yinon and Zitrin, 1996.[10]

1.6.8 Triggering Mechanisms for Explosive Devices

Since explosives can use a large variety of triggering mechanisms, it is critical that emergency responders leave suspicious looking objects untouched and contact appropriate bomb disposal authorities. The mechanisms can be triggered by vibrations, barometric pressure changes, temperature changes, magnetic or photoelectric switches, antiopening switches, timing switches, or radio-controlled switches. For additional details on triggering mechanisms, refer to Pickett, M., 1999.[12]

REFERENCES

1. Siegrist, D.M. and Graham, J.M., *Countering Biological Terrorism in the U.S.: An Understanding of Issues and Status*, Oceana Publications, Dobbs Ferry, NY, 1999.
2. Gosling, F.G., *The Manhattan Project: Making the Atomic Bomb*, U.S. Department of Energy, Washington, D.C., 1999.
3. Rhodes, R., *Dark Sun: The Making of the Hydrogen Bomb*, Simon & Schuster, New York, 1995.
4. Rhodes, R., *The Making of the Atomic Bomb*, Simon & Schuster, New York, 1986.
5. Schwartz, S.I., *Atomic Audit: The Costs and Consequences of U.S. Nuclear Weapons since 1940*, Brookings Institution Press, Washington, D.C., 1998.
6. Cochran, T.B., Arkin, W.M., and Hoenig, M.M., *Nuclear Weapons Databook: Volume I, U.S. Nuclear Forces and Capabilities*, Ballinger, Cambridge, 1984.
7. Caldicott, H., *The New Nuclear Danger*, New Press, New York, 2002.
8. Akhavan, J., *The Chemistry of Explosives*, Royal Society of Chemistry, Letchworth, U.K., 1998.
9. Suceska, M., *Test Methods for Explosives*, Springer-Verlag, New York, 1995.
10. Yinon, J. and Zitrin, S., *Modern Methods and Applications in Analysis of Explosives*, John Wiley & Sons, New York, 1996.
11. Dobratz, B.M. and Crawford, P.C., *LLNL Explosives Handbook: Properties of Chemical Explosives and Explosive Simulants*, Lawrence Livermore National Laboratory, Livermore, CA, 1985.
12. Pickett, M., *Explosives Identification Guide*, Delmar Publishers, New York, 1999.

2 General Types of Radiation and Warfare Agents

This chapter summarizes the various types of radiation and warfare agents that emergency responders and the public may encounter as a result of a terrorist attack. It also discusses recent black market activities, to emphasize the real threat of terrorist actions involving radiological weapons. While the chemical and biological warfare agents presented in Sections 2.2 and 2.3 are the most likely to be encountered, new agents are continuously developed.

2.1 RADIATION

Radioactivity results when some part of an atom is unstable. The instability exists because the orbital electrons or the nucleus contain too much energy. Radioactive atoms are called radionuclides. They release excess energy by emitting radiation. The type of radiation released (alpha, beta, or gamma particles)* may be more or less hazardous to humans, depending on the location of the radioactive materials. Exposure to radioactive materials outside the body poses external hazards. Radioactive materials may also be hazardous when ingested, inhaled, or injected and thus pose internal hazards. The sections below describe the characteristics of radiation particles as external or internal hazards and as they may be encountered after a terrorist attack. Chapter 3 provides additional details and addresses health effects associated with exposure to radiation.

2.1.1 External Radiation

Two radiation particles are considered external radiation hazards: gamma particles and beta particles. Although gamma particles are considered more significant, beta-emitting and gamma-emitting radionuclides can produce large radiation doses,** especially when individuals are exposed to radiological weapons. Alpha particles alone

* Other types of radiation exist, but for the purposes of simplicity, only alpha, beta, and gamma particles are discussed. The term "particle" is used even though radiation may be defined differently under other applications.

** "Dose" as defined here refers to the absorption of radiation energy by human tissue. Higher doses correspond to higher potential for adverse health effects, as described in Chapter 3. The terms "dose" and "exposure" are often used interchangeably.

do not present external radiation hazards, although alpha-emitting radionuclides may also emit gamma particles. A detailed discussion on ways to reduce exposure to external radiation is presented in Chapter 4. Three general safety factors inherent in radiation protection philosophy are also cited: *time, distance,* and *shielding*.

2.1.1.1 Gamma Particles

It is likely that radiological materials encountered after a terrorist attack will emit gamma particles. Gamma particles are photons (particles of light) similar to light that emanates from an ordinary bulb. The difference is that the gamma particles are highly energetic and can damage human tissue. As with ordinary light, gamma particles can travel long distances and cause harm even if an individual is not in direct contact with the radioactive material.

The most extreme case of gamma radiation dose would arise from explosion of a nuclear weapon. Nuclear weapons release intense gamma radiation that can produce fatal doses miles from an explosion (see Chapter 5). A less extreme but more likely scenario involves radioactive materials dispersed via conventional explosives (dirty bombs), where only the immediate area is contaminated with gamma-emitting radionuclides.

Gamma particles may travel great distances in the open air but, as with ordinary light, may also be shielded by wood, bricks, or other solid materials. Consider a lamp in the corner of a room. As an individual walks closer to the lamp, the intensity of the light increases. He may also notice that the light reflects off the walls, increasing the level of brightness. As the individual leaves the room and travels down a hall, the intensity of the light drops significantly especially after turning a corner or moving behind a solid material like a door.

If the lamp were replaced by a source of gamma-emitting radionuclides, gamma radiation levels would increase as the individual approached the source. Some gammas would bounce off the walls, increasing the level of exposure. As the individual left the room, turned a corner, and increased his distance from the source, the walls and other solid materials would act as radiation shields and the level of radiation intensity (dose) would decrease significantly. As indicated by this example, quickly increasing the distance from the source and seeking shelter behind solid material can significantly reduce external exposure to gamma-emitting radionuclides.

Examples of gamma-emitting radionuclides that may be used in a terrorist attack include cobalt-60, cesium-137, radium-226, and others. Nuclear fuel used by nuclear power plants to generate energy also contains a significant number of gamma-emitting radionuclides. Spent (used) fuel and associated waste products are particularly hazardous because of the intense gamma radiation they emit.

2.1.1.2 Beta Particles

Radionuclides used in a terrorist attack may emit beta particles. A beta particle is, in effect, an energetic electron ejected from an unstable nucleus. Beta particles can travel several feet in air and damage exposed tissues such as the skin and eyes. The primary external threat from beta-emitting radionuclides is through direct contact. Thus, it is important to move away from contaminated areas and wash contaminated body parts as soon as possible. Although beta particles may damage exposed tissue, they are more likely to be harmful when beta-emitting radionuclides are ingested, inhaled, or injected (e.g., through a puncture wound) because thin layers of clothing, paper, aluminum, wood, and even quantities of air can effectively shield beta particles.

Beta-emitting radionuclides may also emit gamma radiation, so moving outside the range of beta radiation (several feet) may not be enough to eliminate all external exposures. For example, cobalt-60 and cesium-137 emit both beta and gamma radiation. Spent nuclear fuel also contains a significant number of beta-emitting radionuclides.

2.1.1.3 Alpha Particles

Radionuclides that do not emit beta particles likely emit alpha particles. An alpha particle is, in effect, a helium atom (two protons and two neutrons) ejected from an unstable nucleus. An alpha particle can only travel a few inches in air and cannot penetrate the outer layers of dead skin cells. Therefore, alpha particles are not external hazards and produce tissue damage only if alpha-emitting radionuclides are ingested, inhaled, or injected.

Alpha-emitting radionuclides may also emit gamma radiation, so moving outside the range of alpha radiation (a few inches) may not be enough to eliminate all external exposures. For example, radium-226 emits both alpha and gamma radiation. Other alpha emitters that could be used in a terrorist attack include uranium and plutonium, among others. Spent and fresh (unused) nuclear fuels contain significant numbers of alpha-emitting radionuclides.

2.1.2 Internal Radiation

Unlike external radiation, gamma, beta, *and* alpha particles are all internal hazards. In fact, alpha particles (or alpha-emitting radionuclides) pose the largest internal hazard of the three types of radiation for two reasons. First, an alpha particle inside the body does not have to penetrate skin or other protective layers to impact sensitive tissues and organs. Second, alpha particles are significantly more massive than beta or gamma particles and can produce significant tissue damage over a short range.

Consider, for example a gun that can shoot BBs, golf balls, and bowling balls with the same energy. BBs represent gamma particles, golf balls represent beta particles, and bowling balls represent alpha particles. BBs (gamma particles) can travel long distances but produce only localized damage after hitting a target. Golf balls (beta particles) cannot travel as far but produce more widespread damage when they hit targets at relatively close ranges. A bowling ball cannot travel far but can produce massive damage when it hits a target at close range. Alpha particles emitted from ingested, inhaled, or injected radionuclides no longer have to pass through protective barriers (e.g., skin or clothing) and are in range of sensitive cells such as lungs and bone marrow. For this reason, it is important to limit internal exposure to radiological contaminants, especially alpha-emitting radionuclides.

If threatened by radiological contaminants after a terrorist attack, an individual can use a folded handkerchief, dust mask, gas mask, or any available material to limit the ingestion or inhalation of contaminants. Potentially contaminated puncture wounds can be washed to limit internal doses, although medical treatment by a trained professional is advised, as described in Chapter 7. A more detailed discussion on reducing exposure to internal radiation is presented in Chapter 4.

2.1.3 Radiation Source Material

This section describes specific radiological materials that could be used in a terrorist attack. It also tabulates and describes recent black market activities involving radiological materials.

2.1.3.1 Potential Source Materials

For the purpose of this discussion, radiological materials that could be used in a terrorist attack are divided into three categories: (1) bomb-grade nuclear material, (2) nuclear reactor fuel and associated waste products, and (3) industrial sources. Bomb-grade nuclear material includes concentrated plutonium and/or highly enriched uranium* (>20% U-235) that may be used to build a nuclear weapon, assuming a terrorist group cannot or has not already secured an assembled weapon.

Nuclear fuel and associated waste products also include plutonium and enriched uranium (<20% U-235) and associated waste or fission products that emit intense radiation and can pose significant threats if dispersed with conventional explosives (i.e., by a dirty bomb). Industrial sources include a range of devices used in geological investigation and radiography, and may also pose significant hazards if dispersed by a dirty bomb. Examples of radioactive materials that could be used in a dirty bomb include:

- Nuclear fuel from a nuclear power plant
- Small radioactive sources used by the nuclear medicine industry
- Large medical radioactive sources used for radiography and teletherapy
- Radioactive devices used for food irradiation in industry
- Radioactive components of instruments used to find cracks in pipelines and aircraft
- Radioactive components used in geological instruments

Table 2.1 lists specific radionuclides that may be present in nuclear fuel rods or industrial sources used to construct a dirty bomb. It also lists the radiological half-lives** of each radionuclide, whether they are present in fresh or spent fuel rods, and their potential industrial applications. Note that the actual suites of isotopes for given fuel rods will vary depending on the origin and composition of the original fuel mixture. The uranium and plutonium isotopes found in fuel rods may also be

* "Highly enriched uranium" is uranium composed of >20% by weight of the U-235 isotope.

** "Half-life" is the time it takes a radioactive material to release half of its radioactive energy. Half-life may also be defined as the time required for radioactivity levels to drop by a factor of 2.

found in bomb-grade nuclear materials. The radionuclides in the spent rod category in the table are limited to those with half-lives over 6 months. The 6-month cutoff was used to focus on the most likely contaminants, as most radionuclides with half-lives shorter than 6 months will have decayed away, thus posing little to no threat.

Radiation activity levels for a dirty bomb made with spent fuel depend on the age of the fuel. A simple rule to consider is that any radionuclide will decay to 1% of the original concentration after seven half-lives or will decay to insignificant concentrations after ten half-lives. Therefore, if a fuel rod is removed from a reactor several days before detonation in a dirty bomb, all the isotopes listed in Table 2.1 will likely be present. If a fuel rod was last used in a reactor 5 years before detonating a bomb, ruthenium-106 and cerium-144 will have decayed to insignificant concentrations.

2.1.3.2 Black Market Activities involving Radiological Materials

Ewell (1998)[1] reports that between 1992 and 1994, in at least seven cases, highly enriched uranium and/or plutonium were stolen from the former Soviet Union with the intent of trading it on the black market (see Table 2.2). The first case involved the theft of 1.5 kilograms of highly enriched uranium (90% U-235) from the Luch Scientific Production Association, a Russian nuclear research facility located outside Moscow. Fortunately, on October 9, 1992, the Russian police intercepted the material before it was taken out of the country. In another case, two men stole 4.5 kilograms of highly enriched uranium (20% U-235) from a naval shipyard in Sevmorput, Russia. They were turned over to authorities in June 1994 as they attempted to find a buyer for the material. Table 2.2 details these and five other cases involving the theft of highly enriched uranium and/or plutonium.

The incidents cited in Table 2.2 suggest that it is possible that one or more terrorist groups could gain access to sufficient quantities of highly enriched uranium and/or plutonium to build a nuclear weapon. However, gaining access to the materials is only the first of many complicated steps required to build a nuclear weapon. For this reason, it is far more likely that a terrorist organization would construct a more primitive weapon utilizing highly enriched uranium and/or plutonium or more accessible radioactive materials combined with conventional explosives such as those listed in Table 2.1. Between January 1995 and December 1997, about three dozen reported cases involved smuggling of low-enriched and natural uranium. The amounts recovered, if any, are unknown and any combination of these materials could be used in a dirty bomb.

Smuggled nuclear materials can be purchased on the black market by terrorist organizations. Cases in which terrorists gained access to nuclear fuel have been documented. In 1998, the International Atomic Energy Agency expressed concern that a nuclear fuel rod disappeared from the Congo's University of Kinshasa. The rod was later recovered when terrorists attempted to smuggle it into Italy. On July 4, 2002, a second nuclear fuel rod was reported missing at a research reactor in the

TABLE 2.1
Radionuclides in Nuclear Fuel and Industrial Source

Radioisotope		Half Life		Primary	Nuclear Fuel Constituent		Industrial
Full Name	Abbreviated Form			Radiation [a]	Fresh Rod [b]	Spent Rod [c]	Application [d]
Cobalt-60	Co-60	5.27	Years	Beta/gamma		X	A, B, C
Strontium-90	Sr-90	29.1	Years	Beta		X	C
Ruthenium-106	Ru-106	368	Days	Beta		X	
Cesium-134	Cs-134	2.06	Years	Beta/gamma		X	
Cesium-137	Cs-137	30	Years	Beta/gamma		X	A, B, C
Cerium-144	Ce-144	284	Days	Beta/gamma		X	C
Iridium-192	Ir-192	74	Days	Beta/gamma			A
Uranium-233	U-233	1.59E + 05	Years	Alpha		X	
Uranium-234	U-234	2.45E + 05	Years	Alpha	X	X	
Uranium-235	U-235	7.04E + 08	Years	Alpha/gamma	X	X	
Uranium-236	U-236	2.34E + 07	Years	Alpha	X	X	
Uranium-238	U-238	4.47E + 09	Years	Alpha	X	X	
Plutonium-238	Pu-238	87.7	Years	Alpha	X	X	B
Plutonium-239	Pu-239	24100	Years	Alpha	X	X	B
Plutonium-240	Pu-240	6540	Years	Alpha	X	X	B
Plutonium-241	Pu-241	14.4	Years	Beta	X	X	B
Plutonium-242	Pu-242	3.76E + 05	Years	Alpha		X	B
Americium-241	Am-241	432	Years	Alpha/gamma		X	B, D
Curium-244	Cm-244	18.1	Years	Alpha		X	
Californium-252	Cf-252	2.64	Years	Alpha			C

[a] Gamma listed if considered a significant contributor to exposure.
[b] Potential components of fuel rod prior to use in a reactor.
[c] Nongaseous radionuclides with half lives greater than 6 months; *Reactor Safety Study*, Appendix VI to WASH-1400, October 1975, Table VI 3–1.
[d] Adapted from EPA-520/4–73–002 1974. A = industrial radiography source; B = borehole logging source; C = radiation gauges, automatic weighing equipment; D = smoke detector.

TABLE 2.2
Cases Involving Thefts of Highly Enriched Uranium and Plutonium

Case Name	Material Diverted	Origin of Material	Date Material Recovered	Court Ruling
Podolsk	1.5 kg 90% highly enriched uranium	Luch Scientific Production Association, Podolsk	October 9, 1992: Russian police operation intercepted the smuggler, Leoni Smirnov, in the Podolsk train station	March 11, 1993: Smirnov received 3 years' probation for stealing and storing radioactive material
Andreeva Guba	1.8 kg 36% highly enriched uranium	Naval base storage facility, Andreeva Guba	July 29, 1993: Russian security force arrested the suspects before they could smuggle the material out of Russia	November 2, 1995: Antonov and Popov were sentenced to prison (4 and 5 years, respectively)
Tengen	6.15 g Plutonium-239	Uncertain; possibly Arzamas 16	May 10, 1994: Police stumbled on a cache containing Pu-239 while searching Adolf Jaeckle's apartment for another reason	November 23, 1995: Jaeckle was sentenced to 2.5 years in prison for illegal possession of fissile material
Landshut	800 mg 87.7% highly enriched uranium	Uncertain; possibly Obninsk	June 13, 1994: Undercover German police acted as potential customers in sting operation	Gustav Illich and Vaclav Havlik received prison terms of 19 and 13 months, respectively
Sevmorput	4.5 kg 20% highly enriched uranium	Naval shipyard, Sevmorput	June, 1994: Brother of a suspect asked a coworker for help finding a customer for the material; co-worker notified authorities	Aleksei Tikhomirov and Oleg Baranov were sentenced to 3.5 years in prison.

-- continued

TABLE 2.2 (continued)
Cases Involving Thefts of Highly Enriched Uranium and Plutonium

Case Name	Material Diverted	Origin of Material	Date Material Recovered	Court Ruling
Munich	560 g radioactive fuel; 363 g Plutonium-239	Uncertain; possibly Obninsk	August 10, 1994: Undercover German police acted as potential customers in sting operation	April 24, 1997: Justiniano Torres, Julio Oroz, and Javier Bengoechea were sentenced to prison for 4 years 10 months, 3 years 3 months, and 3 years 9 months respectively
Prague	2.7 kg 87.7% highly enriched uranium	Uncertain; possibly Obninsk	December 14, 1994: Anonymous tip received about material's location	October 1997: Four individuals sentenced for 3 to 9 years in prison

Source: Ewell, E.S., NIS nuclear smuggling since 1995: a lull in significant cases? Nonproliferation Rev. Spring–Summer 1998. With permission.

Democratic Republic of Congo. It contained low-enriched uranium with a U-235 content of 19.9% — just below the 20% threshold defining highly enriched uranium. It is uncertain whether the rod was ever recovered.

On January 10, 1999, the *Philadelphia Inquirer* reported that nuclear fuel was stolen from the Ignalina nuclear power station in Lithuania and was buried in a nearby forest. Authorities later learned that the theft occurred in 1992 when four men stole a 20-kilogram uranium fuel rod and buried it. In November 1997, three former soldiers in the security battalion guarding Ignalina were convicted of stealing the rod with intent to sell it. The *Taipei Times* reported on March 27, 2002 that a radioactive fuel rod was found at a steel plant in Kaohsiung, Taiwan. The article reported that the U.S. Atomic Energy Commission was also searching for five other radioactive rods lost earlier that year.

2.2 CHEMICAL WARFARE AGENTS

Chemical warfare agents generally fall into four major categories: (1) blister agents, (2) blood agents, (3) choking agents, and (4) nerve agents.

Blister agents were developed for military purposes and are intended to inflict casualties (delayed following exposure), restrict terrain access, and slow troop movement. They affect the eyes and lungs and blister the skin. Mustard was the primary blister agent used in World War I. It was recognized by its distinctive odor. Modern blister agents are odorless. Most inflict little or no pain at the time of exposure, but lewisite (L) and phosgene oxime (CX) cause immediate pain upon contact. The onset of symptoms is delayed. Blister agents are divided into five categories, presented here in the order in which they were developed:

- Mustards: distilled mustard (HD)
- Arsenicals: methyldichloroarsine (MD), phenyldichloroarsine (PD), ethyldichloroarsine (ED), and lewisite (L)
- Nitrogen mustards: 2,two-dimensionalichlorotriethylamine (HN-1), 2,dichloro-N-methyldiethylamine (HN-2), and 2,2,2-trichlorotriethylamine (HN-3)
- Oximes: phosgene oxime (CX)
- Mixes: distilled mustard (HD) combined with lewisite (L) and distilled mustard (HD) combined with bis (2-chloroethyl sulfide) monoxide (T)

Blood agents are absorbed into the body primarily through breathing. They affect the body by preventing normal utilization of oxygen by cells. This causes rapid damage to body tissues. Three of the most common blood agents are hydrogen cyanide (AC), cyanogen chloride (CK), and arsine (SA).

Choking agents are designed to target the respiratory tract. They cause the lungs to fill with liquid and cause death by dry land drowning. The primary choking agents are chlorine, phosgene, and diphosgene.

While nerve agents vary in molecular structure, they all exert the same physiological effect on the body: an increase in acetylcholine throughout the body caused by interference with a vital enzyme known as cholinesterase. The four primary nerve agents are tabun, sarin, soman, and VX.

The routes by which one may be exposed to chemical agents vary and include inhalation, ingestion, skin absorption, and local skin and eye impacts. Personnel exposed to any of these agents should be decontaminated immediately following the procedures outlined in Section 7.3 and given immediate medical attention. Refer to Chapter 3 for details about individual chemical agents including:

- Standard NATO code
- Chemical name
- Chemical formula
- Vapor density
- Flash point
- Concentrations that are immediately dangerous to life or health (IDLH)
- Concentrations in the air that will kill 50% of those exposed via inhalation (LC_{50})
- The amount of liquid or solid material required to kill 50% of those exposed via the skin (LD_{50})
- Persistence
- Color
- Odor

- Routes of exposure
- Symptoms of exposure
- Body organs impacted
- Downwind evacuation distances
- Treatment options

2.3 BIOLOGICAL WARFARE AGENTS

Hundreds of biological agents are candidates for terrorist weapons; however, some are more likely than others. The U.S. Department of Defense published a list of the 17 most likely biological agents that fall into three categories:

- Bacteria (anthrax, plague germs)
- Viruses (smallpox, encephalitis, and hemorrhagic fevers like Ebola, Lassa and Rift Valley fevers)
- Toxins that attack the central nervous system (botulinum, fungal toxins, and ricin).

Table 2.3 presents a list of biological agents in each of these three categories. While it shows that vaccines are available for many of these agents, the vaccines

TABLE 2.3
U.S. Department of Defense's 17 Most Likely Biological Weapons

Category	Specific Name	Vaccine Available
Bacterial	*Bacillus anthracis* = anthrax	Yes
	Yersinia pestis = plague	Yes
	Brucella melitensis = brucellosis	Yes
	Francisella tularensis = tularemia	Yes
	Coxiella burnetii = Q fever	Yes
	Vibrio cholerae = cholera	Yes
	Burkholdera pseudomallei = melioidosis	No
Viral	*Variola major* = smallpox	Yes
	Venezuelan equine encephalitis = encephalitis	Yes
	Crimean Congo fever = hemorrhagic fever	Yes
	Rift Valley fever = hemorrhagic fever	Yes
Toxin	*Clostridium botulinum* = botulism toxins	Yes
	Clostridium perfringens toxins	No
	Staphylococcus enterotoxin B/TSST-1	Yes (experimental)
	Ricin	Yes (experimental)
	Saxitoxins	No
	Trichothecene mycotoxins (T-2)	No

have been used only to protect soldiers sent to other countries, counterterrorism units, and emergency responders.

Since symptoms may not surface for days or even weeks following exposure, one of the most difficult challenges for emergency responders and health care professionals is determining the biological agents involved. Even after an unusual pattern of illness begins to emerge, doctors have the difficult task of differentiating terrorism-related illnesses from those that have natural causes. The greatest concern is that lives could be lost if a diagnosis takes too long.

The time lag between exposure to a biological agent and the appearance of symptoms may serve as both a plus and a minus for terrorists. An incubation period allows terrorists time to attack and escape before an alarm is sounded. Conversely, an extended incubation period may diffuse the impact that the terrorists hope to achieve to bring attention to their causes (Tierno, 2002).[2] The routes of exposure to biological agents vary, but may include:

- Inhalation
- Ingestion
- Direct contact with skin breaks or abrasions
- Direct contact with skin lesions or secretions
- Contact with infected persons
- Injection
- Bites from infected flies, fleas, ticks, or mosquitos

The mechanisms of dispersion of biological agents of greatest concern are those that have the potential for impacting the largest populations. These mechanisms include dispersing biological agents in aerosol form over highly populated areas using sprayers or crop dusters, introducing biological agents to urban building ventilation systems, adding biological agents to drinking water sources or food supplies, sending agents through the mail, and a number of other methods.

Refer to Chapter 3 for specific details about individual biological agents including diseases caused, variations of the diseases, incubation periods, routes of exposure, transmissibility (person to person or otherwise), available vaccines, symptoms of exposure, and treatment methods.

REFERENCES

1. Ewell, E.S., NIS nuclear smuggling since 1995: a lull in significant cases? *Nonproliferation Rev.*, Spring–Summer 1998.
2. Tierno, P.M., Jr., *Protect Yourself against Bioterrorism,* Pocket Books, New York, 2002.

3 General Hazards from Exposure to Radiation and Warfare Agents

While the guidance provided in this book is designed to minimize exposure to weapons of mass destruction, it will not eliminate the possibility for exposure. For this reason, it is important to understand the general characteristics of nuclear, chemical, and biological agents, the symptoms of exposure, and potential treatment options. For additional details on chemical structure, chemical characteristics, pathology, and field behavior, see References 1 through 3.

This book does not recommend that anyone use medications or vaccines without first consulting a physician. The authors do not suggest that readers attempt to obtain these medications or vaccines, because the chance of exposure to the hazards cited is small. The material is provided for information purposes only.

3.1 RADIATION

Health effects from exposure to radiation fall into two categories: stochastic (based on probability) and acute. Stochastic effects typically take several years to materialize (e.g., cancer appearing 20 years after an exposure) while acute effects such as nausea or reddening of the skin may take only weeks, days, or even hours to materialize. Stochastic and acute effects are described in more detail in the following sections. First, however, a brief discussion describes how radiation damages human tissue and why exposure may produce one or a combination of the described health effects.

3.1.1 Radiation Damage in Human Tissue

Radiation particles produce damage on the atomic scale primarily by creating ions and breaking chemical bonds. An atom consists of a positively charged nucleus and negatively charged electrons. Ions are created when radiation displaces electrons from an atom, leaving free-floating negatively charged electrons and positively charged atoms. Because alpha, beta, and gamma particles (see Section 2.1) displace electrons and create ions, the term *ionizing radiation* is used, and the displaced electron and positively charged atom are called an *ion pair.*

Ion pairs created by ionizing radiation eventually produce free radicals that disrupt the biochemistry of cells, break chemical bonds, and otherwise produce cell damage. Free radicals are highly reactive atoms that scavenge electrons from other atoms or molecules, causing a chain reaction that can produce cell and tissue damage.

A radiation particle may also break chemical bonds through a direct collision. Thus, a DNA strand or other cellular structure could be damaged directly by a radiation particle or through an interaction with a free radical. In either case, the damage may lead to cell death or mutation. If the cell death count is significant in a particular tissue or organ, its function may be disrupted (e.g., skin reddening) or even cease (e.g., necrosis). However, the short-term function of a tissue or organ is only a relevant factor for large acute doses as discussed in Section 3.1.3. Tissue or organ function is likely not a concern when the cell death count is low; because dead cells cannot reproduce, they cannot mutate and lead to cancer.

A mutated cell may reproduce and begin the formation of a carcinogenic mass (tumor), and mutations may occur after acute or chronic exposure. The specific relationship between acute or chronic exposure rate and cancer risk is hotly debated, although current U.S. regulations conservatively adopted the linear no threshold (LNT) model. This model states that risk is linearly proportional to the total dose even at the smallest possible dose levels (risk is associated with all levels of dose no matter how small). An alternate model theorizes that no measurable adverse health effects appear below doses of about 10 to 25 rem (0.1 to 0.25 Sv). Data supporting both models are limited and, to be conservative, levels of exposure should be kept as low as reasonably achievable (ALARA). Victim and emergency responder doses and dose rate may not be easily controlled in the event of a terrorist attack. However, methods to achieve ALARA exposures are described in Chapters 4 and 5.

The human body is equipped to deal with nominal levels of radiation doses. Background (natural) radiation from radon gas, cosmic sources, soil, and water produces an average dose of about 0.3 rem (0.003 Sv) per year.[4] However, large doses of radiation generated after a terrorist attack can overwhelm the body's ability to repair damage, leading to stochastic or acute health effects.

3.1.2 Chronic Radiation Exposures

Stochastic radiation effects are typically associated with those that occur over many months or years (i.e., are typically chronic instead of acute). Chronic doses are typically on the order of background doses (0.3 rem [0.003 Sv] or less) and are not necessarily associated with larger doses that could result from a terrorist attack with radiological weapons. However, stochastic health effects are defined here as effects that occur many years after chronic or acute exposure to radiological contaminants. Stochastic effects are categorized as cancers and hereditary effects. Because no case of hereditary effects (e.g., mutation of future generations) has been documented, this discussion focuses on cancer risk.

A significant body of data defines the relationship between radiation dose and cancer incidence. This dataset is primarily from a study of the atomic bomb survivors from Nagasaki and Hiroshima, Japan but also includes data from animal studies and other sources of information. While additional data are continuously collected and the dose-to-risk models are often revised, the relationship is simplified here for brevity. More detailed information on the dose-to-risk relationship is readily available on the Internet and in public libraries.

The risk of cancer is often calculated for individual radionuclides. The U.S. Environmental Protection Agency (EPA) publishes radionuclide-specific "cancer slope factors" used to estimate the risk of cancer from chronic exposures. Post-attack investigators may use these factors to assess the risk of cancer to victims and emergency personnel. However, even this level of detail is too complex for the purpose of this discussion. The risk of cancer described by the simple dose-to-risk relationship defined in International Commission on Radiological Protection (ICRP) Publication No. 60[5] is more suitable for this discussion. Table 3.1 summarizes the findings of this report for adult workers and the population as a whole. Adult workers may be defined as emergency responders or other working-age adults; the population as a whole includes children and older adults. The cancer risk coefficients in Table 3.1 are applicable to large groups and are not intended to estimate cancer risk for individuals, although the coefficients are often used this way. The proper use of these factors is demonstrated in the following example:

A large group of individuals including children and adults received average doses of 1 rem (0.01 Sv) after detonation of a dirty bomb. Emergency responders who transported and treated victims received average doses of approximately 0.5 rem (0.005 Sv). Based on the coefficients in Table 3.1, the victims incurred approximately 1 rem $\times 1 \times 10^{-4}$ risk/rem) or a 1 in 10,000 increased risk of a nonfatal cancer from this exposure. They also incurred approximately 1 rem $\times 5 \times 10^{-4}$ risk/rem or a 5 in 10,000 increased risk of a fatal cancer, for a total cancer risk of approximately 6 in 10,000. Similarly, the emergency responders incurred a nonfatal cancer risk of 0.5 rem $\times 8 \times 10^{-5}$ risk/rem or approximately 1 in 25,000 and a fatal cancer risk of 0.5 rem $\times 4 \times 10^{-4}$ risk/rem or 1 in 5000, for a total of approximately 1 in 30,000.

Note that the average risk of cancer for all individuals is about 1 in 4. Therefore, even given the high average doses from this attack, it is much more likely that an individual will get cancer from natural causes.

TABLE 3.1
Radiological Dose to Cancer Risk Coefficients

	Cancer Risk per rem (or per Sv)		
Exposed Population	**Fatal Cancer**	**Nonfatal Cancer**	**Total Cancer**
Adult workers	4×10^{-4} (0.04)	8×10^{-5} (0.008)	4.8×10^{-4} (0.048)
Whole population	5×10^{-4} (0.05)	1×10^{-4} (0.01)	6×10^{-4} (0.060)

Source: Adapted from ICRP, *1990 Recommendations of the International Commission on Radiological Protection*, Publication 60, Pergamon Press, New York, 1991. With permission.

3.1.3 Acute Radiation Exposures

In some circumstances such as may be encountered after a terrorist attack, acute doses may be more important than potential chronic effects. More specifically, high doses of radiation may be immediately dangerous to life and health and could lead to severe injury including sickness, irreparable tissue damage, and death, although the more severe effects would likely only be observed after a nuclear explosion. For the purpose of this discussion, acute exposures are defined as those that occur in a relatively short time (over several days or less) and result in a dose of at least 25–35 rad (0.25–0.35 Gy).[6,7]

Acute exposures are also typically associated only with external exposures. While it is conceivable but highly unlikely that inhaled, ingested, or injected radionuclides could produce large doses over a short time, it is much more likely that acute exposures from a terrorist attack (e.g., with a dirty bomb) would come from external gamma radiation.

Specific health effects resulting from an acute dose appear only after the victim exceeds a dose threshold. That is, the health effect will not occur if doses are below the threshold. (Note that this is significantly different from the LNT model used to predict stochastic effects.) After reaching the acute dose threshold, a receptor can experience symptoms of radiation sickness, also called acute radiation syndrome. As shown in Table 3.2, the severity of the symptoms increases with dose, ranging from mild nausea starting around 25–35 rad (0.25–0.35 Gy) to death at doses that reach 300–400 rad (3–4 Gy). Table 3.2 shows that the range of health effects varies by both total dose and time after exposure.

It is unlikely that a victim of a terrorist attack will use the information in Table 3.2. This information may, however, be useful to emergency responders and medical personnel during triage to identify the most highly exposed individuals. The information may also be useful to postattack investigators who must reconstruct the crime scene and estimate doses to all potential victims.

It is also unlikely that the doses associated with a dirty bomb will produce even the milder acute effects. Although the observation of acute radiation syndrome may be unlikely after a dirty bomb explosion, doses should be kept ALARA to limit the potential for acute and stochastic effects. The entire range of acute radiation syndrome effects will be observed after an attack with a nuclear weapon, as described in Chapter 5

3.2 CHEMICAL AGENTS

Chemical agents are generally characterized as blister agents, blood agents, choking agents, and nerve agents. This section provides details on specific chemical agents within each of these categories.

3.2.1 Blister Agents

Blister agents were developed to inflict casualties, restrict terrain access, and slow troop movement. They affect the eyes and lungs and blister the skin. Mustard was

TABLE 3.2
Acute Radiation Syndrome Summary Table

Dose (rad)	Initial Symptoms	Symptom Onset/End	Clinical Remark
0–25	None	Not applicable	Potential for anxiety
25–75	Nausea, mild headache	Onset: 6 h End: 12 h	Mild lymphocyte depression within 24 h
75–125	Transient mild nausea and vomiting in 5–30% of victims	Onset: 3–5 h End: 24 h	Minimal clinical effect Moderate drop in lymphocytes Increased susceptibility to opportunistic pathogens
125–300	Transient mild to moderate nausea and vomiting in 20–70% of victims Mild to moderate fatigue and weakness in 25–60% of victims	Onset: 2–3 h End: 2 days	Dose can be fatal in 5–50% of victims without treatment (lymphocytes indicate severity) Anticipate infection, bleeding, and fever Increased susceptibility to opportunistic pathogens Wounds and burns geometrically increase morbidity and mortality Significant medical attention required for 10–50% of victims
300–500	Transient moderate nausea and vomiting in 50–90% of victims Mild to moderate fatigue and weakness in 80–100% victims Frequent diarrhea	Nausea/vomiting — Onset: 2 h End: 3–4 days Diarrhea — Onset: 10 day End: 2–3 wk	Fatality expected in 50–100% of victims without treatment Moderate to severe loss of lymphocytes Significant medical attention required Increased susceptibility to opportunistic pathogens

— continued

TABLE 3.2 (continued)
Acute Radiation Syndrome Summary Table

500–800	Moderate to severe nausea and vomiting in 50–90% of victims Moderate fatigability and weakness in 80–100% victims Frequent diarrhea	Onset: 1 h End: indeterminate; may proceed directly into GI syndrome	Fatality expected within 3– 6 weeks in majority of victims
800+	Severe nausea, vomiting, fatigue, weakness, dizziness and disorientation Moderate to severe fluid and electrolyte imbalance, hypotension, possible high fever, and sudden vascular collapse Incapacitation at doses > 1000 rad	Onset: < 3 min End: death expected in 2–3 wk	Fatality expected Bone marrow totally depleted within days (transplant may or may not help) High susceptibility to infection (minor wounds may be fatal)

Sources: Adapted from AFRRI, *Medical Management of Radiological Casualties Handbook*, Special Publication 99–2, Armed Forces Radiobiology Research Laboratory, Bethesda, MD, 1999; and Turner, J.E., *Atoms, Radiation, and Radiation Protection*, McGraw-Hill, New York, 1992. With permission.

the primary blister agent used in World War I and has a distinctive odor. Modern agents are odorless. They generally inflict little or no pain at the time of exposure, except for lewisite (L) and phosgene oxime (CX), which cause immediate pain. The development of casualties from blister agents is delayed. The five categories (in the order developed) are mustards, arsenicals, nitrogen mustards, oximes, and mixes.

3.2.1.1 Mustards

The only agent in this category is distilled mustard (HD). Two other mustard agents (Q and T) developed in the past were dropped from production because they were redundant or were superseded by newer agents. Mustards got their name from their unique smell that resembles burning garlic. Personnel exposed to mustards should be decontaminated immediately and given medical attention.

3.2.1.1.1 Distilled Mustard (HD)

Standard NATO code	HD
Chemical name	bis (2-chloroethyl) sulfide
Chemical formula	$(ClCH_2CH_2)_2S$
Vapor density	5.4 times heavier than air
Flash point	105°C (221°F)
IDLH	—
LC_{50}	—
LD_{50}	7 gm/person
Persistence	Liquids may last several days in warm climates to a month in cold climates
Color	Amber brown liquid
Odor	Similar to burning garlic
Routes of exposure	Inhalation, skin absorption, ingestion, local skin and eye impacts

HD vapors are heavier than air and tend to seek lower elevations. HD is slightly soluble in cold water and soluble in most organic solvents. Exposure in any concentration will cause severe choking. Exposure to vapors in low to moderate concentrations will cause temporary blindness and inflammation of the entire respiratory tract. Higher concentrations cause permanent blindness and strip the bronchial tubes of their mucus membrane linings.

HD is locally and systemically corrosive to human tissue. Local effects include immediate inflammation of the tissues around the eyes and pronounced reddening of the skin. If not decontaminated, the reddened skin will ulcerate into watery boils within 4 to 6 hours. If the blisters are crudely broken, they will often reblister.

Some systemic effects that typically result from prolonged exposure may include internal inflammation and blistering of the throat and lungs. This can lead to pulmonary edema (dry land drowning) in which the windpipe clogs from the bottom to the top. Ingestion causes nausea and vomiting. Absorption into the blood will lead to the destruction of white blood cells. Long-term exposure may also lead to bone marrow destruction and subsequent damage to the immune system.[1]

Due to its flash point of 105°C (221°F), HD and its vapors may explode if exposed to fire or munition detonation. As a form of calibration, a downwind evacuation from a 55-gallon spill should be a minimum of 2.1 miles.[2] See Table 3.3 for a summary of the symptoms of exposure and potential medical treatment options.

Treatment — Patients should be decontaminated immediately prior to treatment using the decontamination method presented in Section 7.3.2. No antidotes are known. Treatment consists of symptomatic management of lesions. If a patient inhaled HD but does not display symptoms of an impacted airway, it may still be appropriate to intubate him because laryngeal spasms or edema may make it difficult or impossible later.[2]

3.2.1.2 Arsenicals

An arsenical is a blister agent based around a chloroarsine ($AsCl_3$) molecule in which one of the chlorine atoms is replaced by an organic radical. Arsenicals include methyldichloroarsine (MD), phenyldichloroarsine (PD), ethyldichloroarsine (ED), and lewisite (L). Personnel exposed to any of these agents should be decontaminated immediately and given medical attention.

3.2.1.2.1 Methyldichloroarsine (MD)

Standard NATO code	MD
Chemical name	Methyldichloroarsine
Chemical formula	CH_3AsCl_2
Vapor density	5.5 times as heavy as air
Flash point	No imminent hazard of explosion or fire
IDLH	—
LC_{50}	46 ppm
LD_{50}	—
Persistence	Very short duration in humid climate; can persist up to several hours in dry cold climates
Color	Colorless liquid or vapor
Odor	Similar to rotting fruit
Routes of exposure	Inhalation, skin absorption, ingestion, local skin and eye impacts

MD is irritating to the eyes and nasal membranes even in low concentrations. Exposure of skin to liquid MD will cause blistering within several hours similar to that caused by HD. Exposure to vapor MD does not typically cause blistering, but it will produce severe respiratory pain, severe damage to the membranes of the lungs, and severe eye discomfort. Dry land drowning can occur as the lungs flood with water and mucus. Splashing liquid MD on the eyes will often result in permanent corneal damage.

Liquid MD penetrates skin on contact. Prolonged skin exposure to its arsenic component will lead to systemic damage through bone calcium displacement and subsequent bone marrow destruction. In its traditional form, MD quickly disperses in open terrain but presents a more prolonged hazard in tightly closed buildings, where it concentrates in basements and substructures due to its vapor density.[1]

As a form of calibration, a downwind evacuation from a 55-gallon spill of this agent should be a minimum of 0.6 miles.[2] See Table 3.3 for a summary of the symptoms of exposure and potential medical treatment options.

Treatment — Patients should be decontaminated immediately prior to treatment using the decontamination method presented in Section 7.3.2. British Anti-Lewisite (BAL) dimercaprol antidote will alleviate some effects. It is available as a solution in oil for intramuscular administration to counteract systemic effects. It is not manufactured currently in the forms of skin and eye ointments.[2]

3.2.1.2.2 Phenyldichloroarsine (PD)

Standard NATO code	PD
Chemical name	Phenyldichloroarsine
Chemical formula	$C_6H_5AsCl_2$
Vapor density	7.7 times heavier than air
Flash point	No imminent hazard of explosion or fire
IDLH	—
LC_{50}	29 ppm
LD_{50}	—
Persistence	Several days in cool dry areas; much shorter in wet humid areas
Color	Clear liquid
Odor	Little or no odor
Routes of exposure	Inhalation, skin absorption, ingestion, local skin and eye impacts

PD is a rare but not entirely obsolete agent effective for penetrating and blistering human skin. It damages the corneas of the eyes, nasal membranes, throat, and lungs and also causes vomiting. Prolonged exposure can cause systemic damage through arsenic–calcium replacement and bone marrow destruction. When inhaled, PD is less lethal than HD.

PD was designed for military use in wet environments because it can persist for several days in dry, cool, shady areas. In open terrain, PD is generally effective only as a vomiting agent. However, due to its extreme vapor density, it is highly effective in enclosed areas (particularly basements), tunnels, gullies, and caves. It is highly effective when delivered as an aerosol from an aircraft.[1]

Minimum evacuation distances have not been calculated for PD. See Table 3.3 for a summary of the symptoms of exposure and potential medical treatment options.

Treatment — Patients should be decontaminated immediately prior to treatment using the decontamination method presented in Section 7.3.2. BAL dimercaprol antidote will alleviate some effects. It is available as a solution in oil for intramuscular administration to counteract systemic effects. It is not manufactured currently in the forms of skin and eye ointments.[2]

3.2.1.2.3 Ethyldichloroarsine (ED)

Standard NATO code	ED
Chemical name	Ethyldichloroarsine
Chemical formula	$CH_3CH_2AsCl_2$
Vapor density	6.0 times heavier than air
Flash point	No immediate hazard of fire or explosion
IDLH	—
LC_{50}	42 ppm
LD_{50}	—
Persistence	Generally <1 hour in temperate climates
Color	Colorless to yellowish liquid
Odor	Similar to rotting fruit
Routes of exposure	Inhalation, skin absorption, ingestion, local skin and eye impacts

ED is a rare but not entirely obsolete agent. It is a blistering agent that causes severe discomfort to the eyes and harsh respiratory effects. Exposure to significant quantities over an extended time can cause damage to bone marrow and to the digestive and endocrine systems. Skin exposure initially results in reddening or rash of the skin. Blistering may appear 2 to 4 hours later. Generally, ED does not cause the same severe subcutaneous damage as sulfur-based mustard agents. When inhaled, it actively attacks lung tissue. Lethal doses will cause dry land drowning. The victim's lungs clog with mucus, dead tissue, water, and blood — causing death from simultaneous asphyxiation and blood poisoning. Damage to lung tissue is permanent for those who survive exposure. Exposure of the eyes to ED can cause permanent corneal damage.

In humid or wet climates, ED breaks down rapidly. It is more persistent in shaded desert areas and creates the greatest hazard in buildings and underground structures such as tunnels, caves, utility conduits, and stagnant sewer lines.[1] As a form of calibration, a downwind evacuation from a 55-gallon spill should be a minimum of 1.0 mile.[2] See Table 3.3 for a summary of the symptoms of exposure and potential medical treatment options.

Treatment — Patients should be decontaminated immediately prior to treatment using the decontamination method presented in Section 7.3.2. BAL dimercaprol antidote will alleviate some effects. It is available as a solution in oil for intramuscular administration to counteract systemic effects. It is not manufactured currently in the forms of skin and eye ointments.[2]

3.2.1.2.4 Lewisite (L)

Standard NATO code	L
Chemical name	Dichloro (2-chlorovinyl) arsine
Chemical formula	$ClCH = CHAsCl_2$
Vapor density	7.1 times heavier than air
Flash point	None
IDLH	—
LC_{50}	—
LD_{50}	2.1 gm/person
Persistence	Heavily splashed liquid will persist for up to a day under temperate conditions; persists much longer in snow
Color	Light amber brown in liquid form; vapor is colorless
Odor	Similar to fresh-cut geraniums
Routes of exposure	Inhalation, skin absorption, ingestion, local skin and eye impacts

In pure liquid form, lewisite causes blindness, immediate destruction of lung tissue, and systemic blood poisoning. It is absorbed through the skin like distilled mustard, but is much more toxic to the skin. Skin exposure results in immediate pain; a rash forms within 30 minutes. Severe chemical burns are possible. Blistering of the skin takes up to 13 hours to develop. Lewisite does not dissolve in human sweat. It commingles with sweat, then flows to tender skin areas such as the inner arm, buttocks, and crotch.

Inhalation death occurs within 10 minutes, primarily a result of dry land drowning in which the lungs and throat fill with mucus, blood, and dead tissue, causing asphyxiation. Two milliliters of pure liquid lewisite absorbed by a 150-pound adult by any means would likely be fatal. Those who survive exposure may continue to show symptoms including pulmonary edema, neural disorders, subnormal body temperature, low blood pressure, and permanent damage to the endocrine system for an indefinite period.[1]

As a form of calibration, a downwind evacuation from a 55-gallon spill should be a minimum of 2.1 miles.[2] See Table 3.3 for a summary of the symptoms of exposure and potential medical treatment options.

Treatment — Patients should be decontaminated immediately prior to treatment using the decontamination method presented in Section 7.3.2. BAL dimercaprol antidote will alleviate some effects. It is available as a solution in oil for intramuscular administration to counteract systemic effects. It is not manufactured currently in the forms of skin and eye ointments.[2]

3.2.1.3 Nitrogen Mustards

Nitrogen mustard agents were introduced before the development of nerve agents. Nitrogen mustard agent HN-1 was developed by accident by the German and Czech pharmaceutical industries. It was originally developed to remove warts and kill malignant growths. It later became a military agent. In contrast, HN-2 was designed as a military agent and later became a pharmaceutical. HN-3 was designed as and remains a military agent. Personnel exposed to these agents should be decontaminated immediately and given medical attention.

3.2.1.3.1 2-Dichlorotriethylamine (HN-1)

Standard NATO code	HN-1
Chemical name	2-dichlorotriethylamine
Chemical formula	$CH_3CH_2N(CH_2CH_2Cl)_2$
Vapor density	5.9 times as heavy as air
Flash point	No immediate danger of fire or explosion
IDLH	—
LC_{50}	22 ppm
LD_{50}	—
Persistence	Persists up to one day in moderate climates; shorter persistence in dry, arid climates; may last up to one week in snowy conditions
Color	Pale amber color in liquid form; vapor is colorless
Odor	Fishy or musty odor
Routes of exposure	Inhalation, skin absorption, ingestion, local skin and eye impacts

HN-1 is a cumulative poison; it builds up in the body with every exposure. It is highly irritating to the eyes and throat. A single vapor exposure may not cause blistering of the skin. However, since the human body does not detoxify HN-1 in appreciable quantities, multiple exposures will produce a skin rash in approximately 30 minutes, followed by blistering of the skin 12 or more hours later. During the 12-hour period prior to blistering, progressive irritation of the respiratory tract will likely occur.

When inhaled, HN-1 penetrates the lung–blood barrier to inhibit the formation of new blood cells in the marrow and degrade blood oxidase function. Dry-land drowning syndrome may not be as pronounced as with distilled mustard or the chloroarsines. However, bronchial pneumonia typically occurs in survivors approximately 24 hours after exposure. When ingested, HN-1 attacks the digestive tract. This often results in lesions in the small intestines, degenerative changes to mucous membranes, and the death of living tissue. Severe bloody diarrhea is a common symptom.

HN-1 is approximately six times heavier than air. As a result, it can concentrate in ground depressions or in the basements of buildings, sewer lines, culverts, and other underground spaces.[1] As a form of calibration, a downwind evacuation from a 55-gallon spill should be a minimum of 2.1 miles.[2] See Table 3.3 for a summary of the symptoms of exposure and potential medical treatment options.

Treatment — Patients should be decontaminated immediately prior to treatment using the chemical agent decontamination method presented in Section 7.3.2. No known antidote is available.

3.2.1.3.2 2-Dichloro-N-methyldiethylamine (HN-2)

Standard NATO code	HN-2
Chemical name	2-dichloro-N-methylethylamine
Chemical formula	$CH_3N(CH_2CH_2Cl)_2$
Vapor density	5.4 times as heavy as air
Flash point	No immediate danger of fire or explosion
IDLH	—
LC_{50}	—
LD_{50}	—
Persistence	Liquid may last up to 1 day in moderate weather or 4 to 5 days in extreme cold
Color	Amber brown in liquid form; vapor is colorless
Odor	Resembles lye soap
Routes of exposure	Inhalation, skin absorption, ingestion, local skin and eye impacts

HN-2 is a cumulative poison, meaning it builds up in the body upon each exposure. It is highly irritating to the eyes and throat and can cause blindness in high concentrations. It interferes with hemoglobin functioning in the blood. Its high volatility and vapor pressure enable it to create blistering through vapor absorption exceeding levels of HN-1 and HN-3. Its liquid toxicity between those of HN-1 and HN-3.

After skin is exposed to HN-2 an epidermal rash develops within approximately an hour. If initial exposure is very low, a rash may not develop. As with HN-1, HN-2 exposure is cumulative. If a person receives multiple low-level exposures, a rash will eventually appear. Blistering will begin about 12 hours after the onset of the skin rash. As with other blister agents, great irritation results when HN-2 vapor or liquid mixes with sweat and flows to tender skin areas (e.g., armpits, buttocks, crotch). Pulmonary effects from exposure to HN-2 are not as severe as for distilled mustard. Dry-land drowning syndrome can occur as the lungs flood with mucus, dead tissue, and blood. The victim dies from a combination of asphyxiation and heart failure.

HN-2 is 5.4 times heavier than air. As a result, it can concentrate in ground depressions or in the basements of buildings, sewer lines, culverts, and other underground spaces.[1] As a form of calibration, a downwind evacuation from a 55-gallon spill of this agent should be a minimum of 2.1 miles.[2] See Table 3.3 for a summary of the symptoms of exposure and potential medical treatment options.

Treatment — Patients should be decontaminated immediately prior to treatment using the decontamination method presented in Section 7.3.2. No known antidote is available.

3.2.1.3.3 2,2,2-Trichlorotriethylamine (HN-3)

Standard NATO code	HN-3
Chemical name	2,2,2-trichlorotriethylamine
Chemical formula	$(ClCH_2CH_2)_3N$
Vapor density	7.1 times heavier than air
Flash point	No immediate hazard of fire
IDLH	—
LC_{50}	18 ppm
LD_{50}	0.7 gm/person
Persistence	May last several days under temperate conditions to an undefined time in extreme winter conditions
Color	Pale amber in liquid form; vapor is colorless
Odor	Odorless
Routes of exposure	Inhalation, skin absorption, ingestion, local skin and eye impacts

HN-3 is similar to HN-1 and HN-2, in that it is a cumulative poison that is highly irritating to the eyes and throat. Permanent corneal damage can occur from vapor exposure alone at concentrations around 200 mg/min/m^3. HN-3 does not create sufficient vapor density to cause rash and blistering from a single exposure. Blistering may result from contact with the liquid form, multiple or persistent vapor exposure, or vapor condensation in sweat. A rash will develop from contact with liquid within approximately 1 hour, followed by blistering 6 to 12 hours later. Toxic effects on the eyes are immediate.

HN-3 penetrates the lung–blood barrier easily. It interferes with hemoglobin functioning in the blood and upon reaching the bone marrow will hinder the production of new blood cells. It also destroys white blood cells, causing damage to the immune system. Dry land drowning may occur following lethal dosage absorption. The lungs become clogged with mucus, dead tissue, and blood. The victim dies from asphyxiation and concurrent heart failure. Survivors may suffer permanent damage to the immune and endocrine systems. Damage to the lungs and throat is also permanent.

Since HN-3 is over 7 times heavier than air, it typically concentrates within the substructures of building and on low-lying terrain. HN-3 is 2 to 3 times more persistent than distilled mustard.[1] As a form of calibration, a downwind evacuation from a 55-gallon spill of this agent should be a minimum of 2.1 miles.[2] See Table 3.3 for a summary of the symptoms of exposure and potential medical treatment options.

Treatment — Patients should be decontaminated immediately prior to treatment using the decontamination method presented in Section 7.3.2. No known antidote is available.

3.2.1.4 Oximes

Phosgene oxime is the only member of this class. It is believed to have originated in the Soviet Union where it was a byproduct of research on insecticides for cockroaches. Exposed individuals should be decontaminated immediately and given medical attention.

3.2.1.4.1 Phosgene Oxime (CX)

Standard NATO code	CX
Chemical name	Dichloroform oxime
Chemical formula	$Cl_2C = NOH$
Vapor density	3.9 times heavier than air
Flash point	—
IDLH	—
LC_{50}	69 ppm
LD_{50}	—
Persistence	Unknown
Color	White in solid form; vapor is colorless
Odor	Musty, pepperish
Routes of exposure	Inhalation, ingestion, local skin and eye impacts

CX is very different from other blister agents in that it attacks whatever tissue it contacts, such as skin, muscle, and nerves. Its impact on nerves is intense unremitting pain. When CX comes in contact with skin, it leaves a blanched area (bleached appearance) within 30 seconds where it has been absorbed into the skin. A red rash-like ring immediately forms around the affected area. Within a day, the blanched area and rash will turn dark as broken-down skin pigment pools near the surface. A scab will form over the area in about a week. This scab will fall off approximately 3 weeks later. Immediate death from systemic shock or trauma is possible. Healing may take as long as 1 year in severe cases.

The impacts from inhalation are not known. Since CX dissolves in human sweat, it can flow to tender skin areas of the body (e.g., armpits, buttocks, crotch). Since its effects are instantaneous, even immediate decontamination may do little to ease pain.[1] As a form of calibration, a downwind evacuation from a 55-gallon spill should be a minimum of 2.1 miles.[2] See Table 3.3 for a summary of the symptoms of exposure and potential medical treatment options.

Treatment — Patients should be decontaminated immediately before treatment using the decontamination method presented in Section 7.3.2. No known antidote is available.

3.2.1.5 Mixes

Blister agents are known to be mixed. The two most common mixed blister agents are:

- Mustard–lewisite mixture (HL): distilled mustard (HD) combined with lewisite (L)
- Mustard-T mixture (HT): distilled mustard (HD) combined with bis (2-chloroethyl sulfide) monoxide (T)

Agent T is out of production. It was created in the mid 1930s and used alone in warm climates or as a distilled mustard thickener. It is obsolete as a pure agent, and semi-obsolete as a mixer.

3.2.2 Blood Agents

Blood agents are absorbed into the body primarily through breathing. They affect the body by preventing normal utilization of oxygen by cells and subsequent rapid damage to body tissues. Three of the most common blood agents are hydrogen cyanide (AC), cyanogen chloride (CK), and arsine (SA). Personnel exposed to these agents should be decontaminated immediately and given medical attention.

3.2.2.1 Hydrogen Cyanide (AC)

Standard NATO code	AC
Chemical name	Hydrogen cyanide
Chemical formula	HCN
Vapor density	0.9 times the weight of air
Flash point	0°F. AC ignites 50% of the time when fired in artillery munitions
IDLH	50 ppm
LC_{50}	180 ppm
LD_{50}	7 gm/person
Persistence	Vapor persistency is very short because the vapor rapidly dissipates; liquid form evaporates at a rate approximately equivalent to water
Color	Colorless in liquid and vapor forms
Odor	Similar to bitter almonds
Routes of exposure	Inhalation, skin absorption, ingestion

AC as a gas can be inhaled or absorbed through breaks in the skin. In liquid form (hydrocyanic acid) it may be ingested, absorbed through the skin or eyes, or inhaled as a mist. It is irritating to the eyes and upper respiratory tract. It inactivates certain enzyme systems, which prevents cells from normal utilization of oxygen.

Low levels of exposure often cause weakness, headache, disorientation, nausea, and vomiting. Higher levels may result in loss of consciousness or may terminate respiration, leading to death within 15 minutes. An immediate lethal dosage often causes violent contractions of blood vessels accompanied by severe shock. This reaction may cause death prior to asphyxiation.[1]

As a form of calibration, a downwind evacuation from a 55-gallon spill should be a minimum of 0.5 miles.[2] See Table 3.3 for a summary of the symptoms of exposure and potential medical treatment options.

Treatment — Patients should be decontaminated immediately prior to treatment using the decontamination method presented in Section 7.3.3. The Lilly Cyanide Antidote Kit contains amyl nitrite, sodium nitrite, and sodium thiosulfate. Dimethylaminophenol, cobalt-edetate, or vitamin B_{12a} are alternative antidotes for cyanide poisoning.[5]

3.2.2.2 Cyanogen Chloride (CK)

Standard NATO code	CK
Chemical name	Cyanogen chloride
Chemical formula	CNCl
Vapor density	2.2 times heavier than air
Flash point	None
IDLH	—
LC_{50}	—
LD_{50}	—
Persistence	Generally short; lethal doses can persist in jungle or forest terrains
Color	Colorless in liquid and gas forms
Odor	Sharp, pepper-like odor similar to those of most tear gases
Routes of exposure	Inhalation, skin absorption, ingestion, and local skin and eye impacts

CK in liquid or gas form is highly irritating to the eyes and upper respiratory tract. Inside the body, it converts to hydrogen cyanide, which inactivates certain enzyme systems that prevent cells from utilizing oxygen. Impacted skin may appear flushed. Low levels of exposure often cause weakness, headache, disorientation, nausea, and vomiting. Higher levels of exposure will result in loss of consciousness, terminate respiration, and cause death within 15 minutes. An immediate lethal dose often causes violent contractions of blood vessels accompanied by severe shock. This reaction may cause death prior to asphyxiation.[1]

As a form of calibration, a downwind evacuation from a 55-gallon spill should be a minimum of 1.3 miles.[2] See Table 3.3 for a summary of the symptoms of exposure and potential medical treatment options.

Treatment — Patients should be decontaminated immediately prior to treatment using the decontamination method presented in Section 7.3.3. The Lilly Cyanide Antidote Kit contains amyl nitrite, sodium nitrite, and sodium thiosulfate. Dimethylaminophenol, cobalt edetate, and vitamin B_{12a} are alternative antidotes for cyanide poisoning.[2]

3.2.2.3 Arsine (SA)

Standard NATO code	SA
Chemical name	Arsenic trihydride
Chemical formula	AsH_3
Vapor density	2.7 times heavier than air
Flash point	—
IDLH	3 ppm
LC_{50}	160 ppm
LD_{50}	—
Persistence	Very short; vaporizes extremely fast; disperses rapidly
Color	Colorless in liquid and vapor forms
Odor	Similar to powdered garlic
Route of exposure	Inhalation

SA is known to accumulate in the body. It displaces calcium in bone matter and impacts the production of new blood cells in bone marrow. SA enters the bloodstream through the respiratory system or breaks in the skin, then damages the liver and kidneys. Skin that has come in contact with SA may appear flushed. Low levels of exposure may cause headache and irritability. Higher levels may cause chills, nausea, and vomiting. Lethal doses may cause blood vessel contraction and shock prior to onset of asphyxiation effects. Death from shock may occur within 2 minutes. Death from asphyxiation may occur within 10 to 15 minutes. Death from kidney or liver damage may occur within 11 days.[1]

As a form of calibration, a downwind evacuation from a 55-gallon spill should be a minimum of 1.5 miles.[2] See Table 3.3 for a summary of the symptoms of exposure and potential medical treatment options.

Treatment — Patients should be decontaminated immediately prior to treatment using the decontamination method presented in Section 7.3.3. No known antidote is available.

3.2.3 Choking Agents

Choking agents are designed to target the respiratory tract. They often cause the lungs to fill with liquid and cause death by dry land drowning. The primary choking agents are chlorine, phosgene, and diphosgene.

3.2.3.1 Chlorine

Standard NATO code	None
Chemical name	Chlorine
Chemical formula	Cl_2
Vapor density	2.4 times heavier than air
Flash point	None
IDLH	10 ppm
LC_{50}	655 ppm
LD_{50}	—
Persistence	Short; rapid dissipation
Color	Greenish-yellow gas
Odor	Similar to concentrated bleach
Routes of exposure	Inhalation, local skin and eye impacts

When inhaled, chlorine strips the linings from the bronchial tubes and lungs. Light inhalation produces vomiting and diarrhea. Heavy inhalation results in immediate inflammation of the bronchial tubes and lungs. Massive amounts of phlegm are formed and hemorrhaging ensues. Dead tissue and liquids fill the lungs, often causing death by dry land drowning. Death may also result from asphyxiation. Survivors typically cough up lung detritus for an extended period. Lung damage is permanent.[1]

As a form of calibration, a downwind evacuation from a 55-gallon spill should be a minimum of 0.5 miles.[2] See Table 3.3 for a summary of the symptoms of exposure and potential medical treatment options.

Treatment — Move patient to fresh air environment and provide oxygen for respiratory distress. Require the patient to rest because even minimal physical exertion may shorten the clinical latency period. No known antidote is available.

3.2.3.2 Phosgene (CG)

Standard NATO code	CG
Chemical name	Carbonyl chloride
Chemical formula	$COCl_2$
Vapor density	3.4 times heavier than air
Flash point	None
IDLH	2 ppm
LC_{50}	79 ppm
LD_{50}	—
Persistence	In an open area, CG persists for only a short duration (~20 minutes); in a sheltered area, it can persist for several days
Color	Colorless gas
Odor	Similar to new-mown hay, grass, or green corn
Route of exposure	Inhalation

CG causes mild irritation to the eyes immediately upon contact. The body does not detoxify CG and it accumulates in the body. CG initially attacks the lung capillaries followed by the membranes of the lung sacs, where it is directly transmitted into the blood and causes the lungs to fill with watery fluids. Lethal doses may not be apparent until 4 hours after exposure. Victims of cumulative doses may take up to a full day before manifesting traumatic symptoms or sudden death.[1]

As a form of calibration, a downwind evacuation from a 55-gallon spill should be a minimum of 1.7 miles.[2] See Table 3.3 for a summary of the symptoms of exposure and potential medical treatment options.

Treatment — Move patient to fresh air and provide oxygen for respiratory distress. Require the patient to rest because even minimal physical exertion may shorten the clinical latency period. No known antidote is available.

3.2.3.3 Diphosgene (DP)

Standard NATO code	DP
Chemical name	Trichloromethyl chloroformate
Chemical formula	$ClC(O)OCCl_3$
Vapor density	6.8 times heavier than air
Flash point	None
IDLH	—
LC_{50}	37 ppm
LD_{50}	—
Persistence	Short duration; more volatile than phosgene; may persist for a day or so in sheltered areas
Color	Colorless oily liquid
Odor	Similar to new-mown hay, grass, or green corn
Routes of exposure	Inhalation

The effects of DP are identical to those of CG (see Section 3.2.3.2). DP causes mild irritation to the eyes immediately upon contact. The body does not detoxify this agent, and it accumulates in the body. DP initially attacks the lung capillaries followed by the membranes of the lung sacs, where it is directly transmitted into the blood and causes the lungs to fill with watery fluids. Lethal doses may not be apparent until 4 hours after exposure. Victims of cumulative doses may take up to a full day before manifesting traumatic symptoms or sudden death.[1]

As a form of calibration, a downwind evacuation from a 55-gallon spill of this agent should be a minimum of 1.7 miles.[2] See Table 3.3 for a summary of the symptoms of exposure and potential medical treatment options.

Treatment — Move patient to fresh air and provide oxygen for respiratory distress. Require the patient to rest because even minimal physical exertion may shorten the clinical latency period. No known antidote is available.

3.2.4 Nerve Agents

While individual nerve agents vary in molecular structure, they all have the same physiological effect on the human body. This effect is an increase in acetylcholine throughout the body caused by interference with a vital enzyme known as cholinesterase. The four primary nerve agents are tabun, sarin, soman, and VX.

3.2.4.1 Tabun (GA)

Standard NATO code	GA
Chemical name	Ethyl N,N-dimethyl phosphoramicocyanidate
Chemical formula	$C_2H_5OP(O)(CN)N(CH_3)_2$
Vapor density	5.6 times heavier than air
Flash point	78°C (172°F)
IDLH	0.03 ppm
LC_{50}	2 ppm
LD_{50}	1 gm/person
Persistence	Evaporates quickly in hot dry climates; splashed liquid may remain for several days in temperatures not exceeding 77°F; may remain for much longer periods in cold weather
Color	Pale to dark amber brown in liquid form; vapor is colorless
Odor	No odor when pure; impurities cause odor of rotting fruit or bitter almonds
Routes of exposure	Inhalation, skin absorption (liquid, vapor), ingestion

GA may be inhaled, absorbed through the skin, or ingested. It is so toxic that contact of the skin with 1.0 to 1.5 grams of liquid solution (~0.01 mg/kg concentration) will cause death within 2 minutes. GA interferes with the neural synapses. It overstimulates the nervous system, which in turn causes over-reactivity of muscles and malfunctioning of various body organs. Autopsies of victims of GA exposure show massive congestion of enzymes and fluids in all the major organs, throughout the nervous system, and most notably in the brain. Those who survive exposure may suffer permanent neurological damage. Detoxification of this agent by the body is very slow.

Due to its density, GA concentrates in lower ground surface elevations, the basements of buildings, and within utility conduits and sewer lines. GA is a liquid that was originally designed to be delivered by artillery munitions or aerial spray.[1]

As a form of calibration, a downwind evacuation from a 55-gallon spill should be a minimum of 5.5 miles.[2] See Table 3.3 for a summary of the symptoms of exposure and potential medical treatment options.

Treatment — Patients should be decontaminated immediately before treatment using the decontamination method presented in Section 7.3.2. The initial treatment for nerve agent exposure is a two-part antidote consisting of atropine and 2-PAM chloride. Atropine is the primary drug for treating nerve agent or organophosphate exposure. It stops the effect of the nerve agent by blocking overstimulation. 2-PAM chloride complements atropine by removing nerve agent at nerve ending sites.

3.2.4.2 Sarin (GB)

Standard NATO code	GB
Chemical name	Isopropyl methyl phosphonofluoridate
Chemical formula	$CH_3P(O)(F)OCH(CH_3)_2$
Vapor density	4.9 times heavier than air
Flash point	None
IDLH	0.03 ppm
LC_{50}	1.2 ppm
LD_{50}	1.7 gm/person
Persistence	Evaporates at approximately the same rate as water, with higher volatility when dispersed as a vapor
Color	Colorless as a liquid or vapor
Odor	No odor in pure state; impurities may have a slight odor similar to rotting fruit
Routes of exposure	Inhalation, skin absorption (liquid or vapor), ingestion

Due to its high vapor density, GB concentrates in lower ground surface elevations, the basements of buildings and within utility conduits and sewer lines. GB is more volatile and toxic than GA and GD. Death occurs rapidly when vapor or liquid is inhaled, ingested, or absorbed through the skin. Direct ingestion of 0.01 mg/kg or more may cause death within 1 minute. The only moderating factor in GB's lethality is its high volatility. Since the body's ability to detoxify GB is very slow, it tends to accumulate in the body.

GB interferes with neural synapses. It causes overstimulation of the nervous system, which in turn causes over-reactivity in the muscles and malfunctioning of various body organs. Exposure to GB will result in massive congestion of enzymes and fluids in all the major organs, throughout the nervous system, and within the brain. Those who survive exposure may suffer permanent neurological damage.[1]

As a form of calibration, a downwind evacuation from a 55-gallon spill should be a minimum of 5.5 miles.[2] See Table 3.3 for a summary of the symptoms of exposure and potential medical treatment options.

Treatment — Patients should be decontaminated immediately prior to treatment using the decontamination method presented in Section 7.3.2. The initial treatment for nerve agent exposure is a two-part antidote consisting of atropine and 2-PAM chloride. Atropine is the primary drug used to treat nerve agent or organophosphate exposure. It stops the effect of the nerve agent by blocking overstimulation. 2-PAM chloride complements atropine by removing nerve agent at nerve ending sites.

3.2.4.3 Soman (GD)

Standard NATO code	GD
Chemical name	Pinacolyl methyl phosphonefluoridate
Chemical formula	$CH_3P(O)(F)OCH(CH_3)C(CH_3)_3$
Vapor density	6.3 times heavier than air
Flash point	250°F
IDLH	0.008 ppm
LC_{50}	0.9 ppm
LD_{50}	0.35 gm/person
Persistence	May last about 2 days in shaded areas under moderate climatic conditions
Color	Colorless as a liquid or vapor
Odor	Similar to that of rotting fruit; impurities may produce the odor of oil of camphor
Routes of exposure	Inhalation, skin absorption (liquid, vapor), ingestion

GD is more neurologically active than GA and GB. A dosage of 0.01 mg/kg (ingestion or skin absorption) may bring death within 1 minute. Since the body's ability to detoxify GD is very slow, it tends to accumulate in the body. Because it persists 1 to 2 days under moderate weather conditions, casualties through dose accumulation are common.

Vapor and liquid forms of GD entering open wounds will cause immediate spasmodic reactions, with death following quickly. Similar to GA and GB, GD interferes with neural synapses. It causes overstimulation of the nervous system, which in turn causes over-reactivity in the muscles and malfunctioning of body organs. Exposure to GB most profoundly affects the brain. Those who survive exposure may suffer permanent systemic damage.[1]

As a form of calibration, a downwind evacuation from a 55-gallon spill should be a minimum of 5.5 miles.[2] See Table 3.3 for a summary of the symptoms of exposure and potential medical treatment options.

Treatment — Patients should be decontaminated immediately prior to treatment using the decontamination method presented in Section 7.3.2. The initial treatment for nerve agent exposure is a two-part antidote consisting of atropine and 2-PAM chloride. Atropine is the primary drug used to treat nerve agent or organophosphate exposure. It stops the effect of the nerve agent by blocking overstimulation. 2-PAM chloride complements atropine by removing nerve agent at nerve ending sites.

3.2.4.4 V-Gas (VX)

Standard NATO code	VX
Chemical name	Ethyl S-2-di-isopropyl aminoethyl methylphosphorothiolate
Chemical formula	$CH_3P(O)(OCH_2CH_3)SCH_2CH_2N(CH_3)_2$
Vapor density	9.2 times heavier than air
Flash point	—
IDLH	—
LC_{50}	—
LD_{50}	—
Persistence	Highly persistent, particularly in acidic soils; may last several weeks in temperate climates
Color	Amber colored oily liquid similar in appearance to 20-weight motor oil; vapor is colorless
Odor	No odor
Routes of exposure	Inhalation, skin absorption (liquid, vapor), ingestion

VX was designed to cause casualties primarily through skin absorption or from inhalation or ingestion of mists from aerial sprays or air-bursting munitions. As little as 0.001 grams entering the body through cuts or scrapes or otherwise through the skin can prove fatal. Since the body's ability to detoxify VX is very slow, it tends to accumulate in the body. It is rapidly absorbed into the bloodstream. Victims who receive moderate doses often exhibit spasmodic symptoms such as jerking, urinating, and defecation prior to death. Those who receive massive doses die almost immediately since their nervous systems, hearts, lungs, and brains immediately cease functioning. Systemic damages are permanent for those who survive exposure.[1]

As a form of calibration, a downwind evacuation from a 55-gallon spill should be a minimum of 5.5 miles.[2] See Table 3.3 for a summary of the symptoms of exposure and potential medical treatment options.

Treatment — Patients should be decontaminated immediately prior to treatment using the decontamination method presented in Section 7.3.2. Ventilate the patient because of a possible increase in airway resistance due to constriction and the presence of secretions. If breathing is difficult, administer oxygen. Administer antidotes as soon as possible. The antidote for this agent is atropine alone or in combination with pralidoxime chloride (2-PAMCl) or another oxime. Diazepam may be required to control severe convulsions.

3.3 BIOLOGICAL AGENTS

Biological agents are generally categorized as bacterial agents, viral agents, or toxins. This section provides details on specific biological agents in these categories.[3]

3.3.1 Bacterial Agents

This section addresses the bacterial forms of biological agents that would most likely be considered by terrorists:

Bacillus anthracis (anthrax)
Yersinia pestis (plague)
Brucella melitensis (brucellosis)
Francisella tularensis (tularemia)
Coxiella burnetii (Q fever)
Vibrio cholerae (cholera)
Burkholdera mallei (glanders)
Burkholdera pseudomallei (melioidosis)[3]

3.3.1.1 *Bacillus anthracis* (Anthrax)

Disease caused	Anthrax
Variations of disease	Inhalation anthrax, cutaneous anthrax, gastrointestinal anthrax
Incubation period	1 to 7 days in most cases; up to 60 days in rare cases for inhalation anthrax
Routes of exposure	Inhalation, ingestion, direct contact with break in skin, fly bite
Person-to-person transmissibility	Only via secretions from cutaneous lesions
Vaccines available	Yes

Depending on the route of exposure to this biological agent, the exposed person may experience inhalation anthrax, cutaneous anthrax, or gastrointestinal anthrax. Any one of these forms can be complicated by meningitis occurring in about 5% of cases when anthrax bacilli enter the central nervous system via the bloodstream and eventually reach the blood–brain barrier.

3.3.1.1.1 Inhalation Anthrax

This form is a biphasic disease in that the initial phase is characterized by mild flu-like symptoms followed by a day or so of apparent wellness immediately followed by the acute phase and more serious symptoms. The incubation period can vary from 1 to 5 days, depending on the number of spores inhaled; in some cases it can be as long as 60 days.

Symptoms during the initial phase are characterized by nonspecific and relatively mild respiratory illness, muscular pain, malaise, fatigue, low-grade fever, and non-productive cough. Patients sometimes complain of mild chest discomfort.

Symptoms of the acute phase are acute respiratory distress, breathing difficulty, profuse sweating, turning bluish in color, high temperature, and increased pulse and respiratory rate with chest sounds. If an x-ray is performed, mediastinal widening (swelling of lymph nodes under the breastbone) is very characteristic. Shock and death usually follow within 24 to 36 hours after the onset of respiratory distress. The fatality rate from inhalation anthrax ranges from 65 to 90% even with antibiotic therapy.[3]

3.3.1.1.2 Cutaneous Anthrax

This skin form of anthrax results after spores are introduced beneath the skin by inoculation or contamination of a preexisting lesion or break in the skin. The incubation period for this form of the disease most often is 2 to 5 days, but can be as long as 7 days. Skin lesions start as small, painless, itchy pimples on some exposed part of the body such as the face, neck, or arms. The lesions become vesicular (like blisters) and small rings of vesicles may develop. The vesicles may join into a single large vesicle that eventually ruptures to form an open ulceration. The ulceration develops a black scab at its center (within 2 to 6 days). The area of the lesion may become swollen. The black scab falls off after 1 to 2 weeks and leaves a permanent scar. Many systemic symptoms such as fever, myalgia, and regional swelling of lymph nodes in the area that drains the site may occur. In some cases, significant swelling of head and neck lymph nodes may interfere with the windpipe. Blood infection is rare but can occur.

Untreated cutaneous anthrax can have a fatality rate of up to 20%. Fatalities are rare when patients are treated with appropriate antibiotics. While anthrax is not transmissible from person to person, direct exposure to vesicle secretions of cutaneous anthrax lesions can produce secondary cutaneous lesions. Secretions from vesicles may be prolific, so caution is advised.[3]

3.3.1.1.3 Gastrointestinal Anthrax

This form of anthrax is caused by ingestion of raw or undercooked contaminated meat. The incubation period for this form is 2 to 7 days. The two types of gastrointestinal anthrax (intestinal and oropharyngeal) have different sets of symptoms.

Intestinal anthrax initially produces nausea, vomiting, loss of appetite, and fever. The disease progresses with increasing abdominal pain and vomiting of blood. Bloody diarrhea can develop and may be accompanied by fluid buildup in the abdominal cavity outside the intestines.

The symptoms of oropharyngeal anthrax include swelling of the neck, and lesions in the oral cavity (similar to cutaneous lesions), sometimes on the tonsils. Other symptoms include fever, swollen lymph nodes, and inability to swallow. Shock and toxemia can characterize both forms of the disease. The fatality rate for gastrointestinal anthrax ranges from 25 to 60%.[3]

Treatment — A number of antibiotics are available for treatment including quinolone antibiotics (ciprofloxacin, levofloxacin, ofloxacin) and tetracycline antibiotics (doxycycline). Penicillin antibiotics (amoxicillin, penicillin V, and penicillin G) are effective only for penicillin-susceptible strains.

Penicillin antibiotics must be used with caution because some strains of *Bacillus anthracis* possess an enzyme that inactivates penicillin. Other antimicrobial agents can be used as alternatives if the listed drugs are unavailable or in short supply. These include erythromycin, imipenem, clindamycin, vancomycin, and chloramphenicol.[3]

3.3.1.2 *Yersinia pestis* (Plague)

Disease caused	Plague
Variations of disease	Bubonic plague, pneumonic plague
Incubation period	2 to 10 days
Routes of exposure	Inhalation, bites by infected fleas
Person-to-person transmissibility	Yes
Vaccines available	Yes

In a biological warfare scenario, the plague bacteria could be delivered by contaminated fleas (bubonic plague) or, more likely, by aerosol spread (pneumonic plague). Pneumonic plague can be transmitted also by large aerosol droplets expelled by coughing.

The incubation period for bubonic plague ranges from 2 to 10 days. Malaise, high fever, and tender lymph nodes (buboes) are characteristic. The enlarged lymph nodes are most often around the groin, but can appear also around the neck region and under the arms. The nodes may be tender and compressible and become necrotic over time. Victims may suffer from the septicemic form of bubonic plague that infects the bloodstream. The bacteria can also spread to the central nervous system, lungs, and elsewhere. Vomiting and diarrhea may develop in early stages. Later, shock, renal failure, and heart failure may occur, leading to death.

The incubation period for pneumonic plague ranges from 1 to 3 days. High fever, cough, and chest pain are prominent features. Production of bloody sputum, headache, chills, malaise, myalgia, and evidence of bronchopneumonia also are characteristic symptoms. As pneumonia progresses, breathing becomes more difficult and the victim begins turning blue. The final result is respiratory and circulatory collapse.[3]

Treatment — A number of antibiotics including tetracycline, streptomycin, gentamicin, chloramphenicol, and quinolone[3] are available for treatment.

3.3.1.3 *Brucella melitensis* (Brucellosis)

Disease caused	Brucellosis
Variations of disease	N/A
Incubation period	1 to 6 weeks
Routes of exposure	Ingestion, direct contact via break in skin, mucous membranes
Person-to-person transmissibility	No
Vaccines available	Yes

Brucellosis is caused by *Brucella melitensis, B. suis, B. abortus, or B. canis* bacteria. After entering the human body, the organisms travel from their entrance point to the lymph channels and nodes, eventually reaching the thoracic duct and bloodstream. Carried through the bloodstream, they are deposited eventually in multiple organs. They grow inside the organs and eventually kill their host cells. A new crop of bacteria is then released.

The incubation period for the bacteria ranges from 1 to 6 weeks, but most commonly is 3 to 4 weeks. The onset of brucellosis is characterized by malaise, fever, chills, sweating, headache, fatigue, myalgia, and arthralgia. Fever accompanied by drenching sweat usually rises in the afternoon and falls during the night. Swollen lymph nodes, spleen, and liver may also be present. The undulant fever can last for weeks, months, or even years. Coughs occur in approximately 20% of cases, while x-rays appear normal. Lethality may approach 6% if *B. melitensis* is the agent and <1% if other species are involved. Most deaths are associated with infection of the lining of the heart (endocarditis) or infection of the membranes around the brain (meningitis). Gastrointestinal symptoms occur in up to 70% of adult cases, and less frequently for children. Rashes occur in <5% of cases.[3]

Treatment — A number of antibiotics are available including doxycycline combined with rifampin, and ofloxacin combined with rifampin. Vaccines for animal use directed against *B. melitensis* and *B. abortus* have proven very successful. An effective vaccine for human use against *B. suis* is currently under development.[3]

3.3.1.4 *Francisella tularensis* (Tularemia)

Disease caused	Tularemia
Variations of disease	Ulceroglandular tularemia, glandular tularemia, typhoidal tularemia, oculoglandular tularemia, oropharyngeal tularemia, and pneumonic tularemia
Incubation period	1 to 21 days
Routes of exposure	Inhalation, ingestion, direct contact with break in skin; bites of infected ticks, mosquitoes, or flies
Person-to-person transmissibility	Yes
Vaccines available	Yes

The six forms of the disease are ulceroglandular tularemia, glandular tularemia, typhoidal tularemia, oculoglandular tularemia, oropharyngeal tularemia, and pneumonic tularemia. The bacterium is highly contagious; as few as 25 inhaled organisms or 10 administered under the skin can cause infection. The common symptoms from infection are:

- Swelling of lymph nodes that remain enlarged for long periods; they eventually become necrotic and drain for long periods
- Fever (usually low)
- Malaise
- Headache
- Pain in regional lymph nodes
- Pneumonia, particularly with exposure to typhoidal tularemia

Ulceroglandular tularemia is the most common form. It comprises 70 to 75% of all cases. Pneumonic tularemia is the next most common (8 to 13%), followed by glandular tularemia (5 to 12%). The other forms are less common. Typhoidal tularemia has the highest mortality rate and as a result is the most likely to be used by terrorists. Exposure causes acute onset of fever, chills, headache, vomiting, and diarrhea. Skin lesions and swollen lymph nodes are not usual. This is a systemic disease and is the only form of tularemia in which diarrhea is usually seen.[3]

Treatment — Doxycycline and tetracycline antibiotics should be considered for treating this disease.[3]

3.3.1.5 *Coxiella burnetii* (Q Fever)

Disease caused	Q Fever
Variations of disease	N/A
Incubation period	2 to 14 days
Routes of exposure	Inhalation, ingestion
Person-to-person transmissibility	Only in rare cases
Vaccines available	Yes

Coxiella burnetii is a highly infectious bacterium. A single bacterial cell can produce clinical illness. For this reason it is a very strong candidate for use by terrorists via aerosol delivery. While the incubation period is 2 to 14 days, the average is 7 days. In rare instances, incubation period can extend up to 1 month. After infection and proliferation in the lungs, the organisms are picked up by macrophages and carried to the lymph nodes, and from there to the bloodstream.

After incubation, initial clinical manifestations include fever, cough, chills, myalgia, headache, and sometimes pleuritic chest pain. Approximately 50% of patients show abnormal chest x-rays; patchy infiltrates resemble viral disease. Uncommon complications include endocarditis, hepatitis, aseptic meningitis, encephalitis, and osteomyelitis. Most patients who develop endocarditis have preexisting valvular heart disease.[3]

Treatment — Various antibiotics are useful in treating *Coxiella* infections. They include tetracycline, doxycycline, and erythromycin. In cases of endocarditis, treatments with doxycycline combined with rifampin, and trimethoprim–sulfamethoxazole combined with doxycycline or tetracycline for 12 months or longer have been successful.[3]

3.3.1.6 *Vibrio cholerae* (Cholera)

Disease caused	Cholera
Variations of disease	Serogroups 01 (classic, El Tor) and 0139
Incubation period	4 hours to 5 days
Route of exposure	Ingestion
Person-to-person transmissibility	Only in rare cases
Vaccines available	Yes

Humans acquire cholera by consuming water or food contaminated with the organism. If terrorists were to use cholera bacteria as a weapon, they would most likely use it to contaminate a water supply. *Vibrio cholerae* produces several toxins, but the cholera toxin is the most important. It causes the mucosal cells of the small intestine to hypersecrete water and electrolytes into the lumen of the gastrointestinal tract. This results in a profuse watery diarrhea that can lead to severe dehydration. Patients become severely hypotensive and can die without medical treatment. The infective dose can be anywhere from 10 to 500 organisms. The incubation period can be as short as 4 hours and as long as 5 days; it is usually 2 to 3 days.

The typical symptoms begin with the sudden onset of nausea and vomiting and profuse diarrhea without abdominal cramps. The stools produced are characteristically "ricewater"-like and contain mucus, epithelial cells, and *ibrio cholerae* bacteria. The dehydration resulting from rapid loss of fluid and electrolytes leads to circulatory collapse and kidney shutdown. Mortality rate without treatment can be as high as 50%.[3]

Treatment — Because of the severe dehydration caused by cholera infection, the most important therapy is fluid and electrolyte replacement. Many antibiotics are effective against *V. cholerae*, including tetracycline, doxycycline, ciprofloxacin, and erythromycin.[3]

3.3.1.7 *Burkholdera mallei* (Glanders)

Disease caused	Glanders
Variations of disease	N/A
Incubation period	3 to 14 days
Routes of exposure	Abrasion of the skin, inhalation
Person-to-person transmissibility	Not likely
Vaccines available	No

Although this disease mainly affects horses, untreated systemic glanders in humans is almost 100% fatal. The primary way that terrorists would weaponize glanders would be to make it an aerosol. The incubation period for the disease ranges from 3 to 14 days, and is dependent upon the sizes of the particles aerosolized and the dose.

Glanders occurs in both acute and chronic forms. The acute form is inhaled and affects the upper respiratory tract. In nature, the acute form kills infected animals in 3 to 4 weeks. The acute form infects the nasal, oral, and conjunctival mucous membranes, causing blood-streaked discharges from the nose along with nodules and ulcerations. The chronic form affects the joints and lymph as multiple skin nodules begin to ulcerate and are filled with pus. Other symptoms include fever, sweats, myalgia, headache, enlarged spleen, and chest pain. Sometimes pneumonia is present. As few as 1 to 10 bacteria delivered to animals by aerosol can be lethal.[3]

Treatment — No vaccines are available for humans. Glanders may be treated with sulfadiazine, doxycycline, rifampin, trimethoprim–sulfamethoxazole, streptomycin, and ciprofloxacin.[3]

3.3.1.8 *Burkholdera pseudomallei* (Melioidosis)

Disease caused	Melioidosis
Variations of disease	N/A
Incubation period	3 to 14 days
Routes of exposure	Ingestion, inhalation, fly bites
Person-to-person transmissibility	Not likely
Vaccines available	No

Melioidosis is very similar to glanders. While it usually causes disease in rodents, it can be transmitted to humans via food contaminated by rodent droppings or biting flies. The primary way for terrorists to weaponize melioidosis would be to make it an aerosol. The mortality rate is 95% in untreated acute disease patients.

The symptoms include pneumonia and fluid around the lungs. The pneumonia may result from inhaling the bacterium during an assault or via bloodstream infection. Any infection with this bacterium may lead to blood infection, which may cause hypotension and shock. The three forms of this disease recognized are acute, subacute, and chronic. The acute form is primarily a bloodstream infection (septicemia). The subacute form mimics tuberculosis, and the chronic form presents an inflammation of skin tissue.[3]

Treatment — No vaccines are available. Melioidosis may be treated with tetracycline, chloramphenicol, trimethoprim–sulfamethoxazole, doxycycline, and ceftazidime.[3]

3.3.2 Viral Agents

This section addresses the viral forms of biological agents that would most likely be considered by terrorists. They include variola major (smallpox), Venezuelan equine encephalitis (encephalitis), Crimean Congo hemorrhagic fever, and Rift Valley hemorrhagic fever.

3.3.2.1 Variola Major (Smallpox)

Disease caused	Smallpox
Variations of virus	N/A
Incubation period	10 to 14 days
Routes of exposure	Inhalation, contact with skin lesions or secretions
Person-to-person transmissibility	Yes
Vaccines available	Yes

The portals of entry for the smallpox virus are the mucous membranes of the upper respiratory tract. Smallpox is transmitted by large or small respiratory droplets and by contact with skin lesions or secretions. Patients are considered more infectious if they are actively coughing. Incubation period ranges from 10 to 14 days, but most typically is 12 days. The only exception is a Russian strain that has a 3- to 5-day incubation period.

Within the first 2 to 3 days of exposure a person will experience symptoms such as malaise, fever, headache, chills, and backache. The fever can last 1 to 5 days. Usually after the fever is gone, a skin eruption or rash appears. It begins as a pimple lesion for 1 to 4 days, becomes blister-like for 1 to 4 days, and then fills with pus for 2 to 6 days. It then forms a crust that falls off 2 to 4 weeks after the first skin lesion appears, leaving pink scars. An important characteristic is that all smallpox lesions in any affected area are generally found in the same state. In contrast, chickenpox lesions are not synchronous; they form in crops. Smallpox lesions are also said to be more numerous on the face and extremities rather than the trunk, unlike chickenpox. The case fatality rate in unvaccinated patients is 15 to 40%. In vaccinated people, the fatality rate is $<1\%$. Patients with smallpox are infectious as soon as a rash is evident and remain infectious until their scabs fall off — a duration of about 3 weeks.[3]

Treatment — Vaccinia immune globulin must be used in conjunction with a vaccinia vaccine if exposure to a smallpox case occurred more than 4 days earlier. However, only the vaccinia vaccine is required less than 4 days after such contact. The vaccine starts to be protective in approximately 7 days. This vaccine does not provide life-long immunity. Revaccination is recommended at 5- to 10-year intervals. Certain antiviral drugs such as Cidofovir have demonstrated that they confer some protection against infection. Unfortunately, because smallpox has been eradicated, limited research on such drugs has been conducted.[3]

3.3.2.2 Venezuelan Equine Encephalitis

Disease caused	Venezuelan equine encephalitis
Variations of virus	Variants A/B and C
Incubation period	4 to 21 days
Routes of exposure	Inhalation, mosquito bites
Person-to-person transmissibility	Moderately transmissible
Vaccines available	Yes

Since this virus has successfully been eradicated from the U.S., any appearance of the virus would likely be sign of a terrorist event. Almost 100% of those exposed to the virus will develop an influenza-like illness. Most encephalitis viruses are destroyed by heat and are easily killed by ordinary disinfectants.

Symptoms of exposure include the rapid onset of fever (usually high), headache, dizziness, lethargy, depression, anorexia, chills, myalgia, photophobia, nausea, vomiting, cough, sore throat; diarrhea may occur. Venezuelan equine encephalitis is not distinguishable from other viruses that cause encephalitis. The acute phase of the disease is 1 to 3 days followed by a prolonged period (up to 2 weeks) of lethargy. Full recovery usually occurs after 2 weeks. Less than 5% of patients would show neurologic manifestations characterized by convulsions, coma, and paralysis during a natural epidemic. The neurologic cases are seen mainly in children; a death rate of up to 20% is possible in children.[3]

Treatment — No specific antiviral therapy exists to treat Venezuelan equine encephalitis. Treatment is geared toward relieving headache or myalgia, controlling convulsions, and aiding breathing difficulties. Several experimental drugs (e.g., vaccines TC-83 and TC-84) have shown some promise in the treatment of this virus, but current data on their effectiveness in humans is insufficient.[3]

3.3.2.3 Crimean Congo Hemorrhagic Fever

Disease caused	Hemorrhagic fevers
Other viral hemorrhagic fevers	Ebola virus, Marburg virus, Lassa fever, Argentine and Bolivian hemorrhagic fevers, Crimean Congo hemorrhagic fever, Rift Valley fever, dengue hemorrhagic fever, yellow fever
Incubation period	4 to 21 days
Routes of exposure	Inhalation, tick bites
Person-to-person transmissibility	In some cases
Vaccines available	Yes, for some strains

Crimean Congo hemorrhagic fever and the other viral hemorrhagic fevers listed above could be weaponized for aerosol delivery. The symptoms of exposure for Crimean Congo hemorrhagic fever are very similar to those of other viral hemorrhagic fevers. The most presenting symptoms of exposure to Crimean Congo hemorrhagic fever are fever, myalgia, low blood pressure, flushing, and black-and-blue marks on the body. Typically, the onset of this virus is 3 to 12 days after tick exposure or inhalation of an aerosol. Patients can suffer extensive gastrointestinal bleeding and ecchymoses. Other symptoms include headache, back pain, nausea, vomiting, and delirium. A jaundiced appearance (yellow coloration of skin and whites of the eyes, evidencing liver failure) and swollen liver are also seen. Mortality from Crimean Congo hemorrhagic fever is 15 to 30%, but some hemorrhagic fevers (e.g., Ebola) have death rates close to 90%.[3]

Treatment — The most important and immediate care needed is the management of hypotension caused by fluid loss. Very aggressive supportive care must be provided. Few specific drugs can be used against the myriad viral hemorrhagic fevers, but ribavarin, an antiviral drug, can be given to Crimean Congo hemorrhagic fever patients. Treatments with immune globulin vaccines have shown usefulness against Crimean Congo hemorrhagic fever, Rift Valley fever, Bolivian hemorrhagic fever, and Lassa fever. Yellow fever vaccine is the only established and licensed vaccine for a hemorrhagic fever. Several vaccines are under development to treat Argentine hemorrhagic fever (live attenuated vaccine), Bolivian hemorrhagic fever (same vaccine), Rift Valley fever (live and inactivated forms). Ribavirin would probably be used for prophylaxis in the event of a suspected or known biological attack.[3]

3.3.2.4 Rift Valley Hemorrhagic Fever

Disease caused	Hemorrhagic fever
Other viral hemorrhagic fevers	Ebola virus, Marburg virus, Lassa fever, Argentine and Bolivian hermorrhagic fevers, Crimean Congo hemorrhagic fever, Congo hemorrhagic fever, dengue hemorrhagic fever, yellow fever
Incubation period	4 to 21 days
Routes of exposure	Inhalation, mosquito bites
Person-to-person transmissibility	Yes
Vaccines available	Yes

Rift Valley fever occurs primarily in sub-Saharan Africa where it is transmitted by mosquitoes. If a terrorist were to use Rift Valley fever, it would be dispersed as an aerosol. One troubling thing about a terrorist attack using this virus is that humans and domestic animals can become infected by aerosol means and a resident mosquito population can continue the assault. Defense against Rift Valley fever must include mosquito control.

Rift Valley fever is characterized by high fever and the development of hemorrhagic areas under the skin. Only a small number of cases (<1%) go on to develop the more serious viral hemorrhagic fever syndrome that causes death in 50% of those who manifest this syndrome. The syndrome is associated with mucosal bleeding or hemorrhaging, liver and kidney failure, and shock before death. Some infections can be complicated with encephalitis and a variety of ocular defects.[3]

Treatment — As with all viral hemorrhagic fevers, supportive therapy must be given, dependent on the complications experienced by patients. Treatment would be intravenous ribavarin for 4 to 6 days. While no human studies to date verify the efficacy of this treatment, cell and rodent studies attest to its efficacy. An effective inactivated vaccine that can be administered in three doses is available. Protective antibodies appear before 14 days and last 1 year. Annual boosters must be given.[3]

3.3.3 Toxins

This section addresses the toxin forms of biological agents that would most likely be considered by terrorists. These agents include *C. botulinum*, *C. perfringens*, *S.* enterotoxin B, ricin, saxitoxins, and trichothecene mycotoxins (T-2).

3.3.3.1 *Clostridium botulinum* toxin

Disease caused	Botulism
Variations of disease	Food-borne botulism, wound botulism, infant botulism, and a fourth undetermined classification; during the growth of the *C. botulinum* bacteria, a variety of toxins (A through G) are produced by different strains; the toxins primarily responsible for human illness are A, B, E, and F
Incubation period	18 to 72 hours
Routes of exposure	Inhalation, ingestion
Person-to-person transmissibility	No
Vaccines available	Yes

C. botulinum toxins cause botulism. They are among the most toxic substances known. A lethal dose for humans is 1 microgram (a millionth of a gram). In a terrorist attack, it is anticipated that *C. botulinum* toxins would be delivered as an aerosol and would produce symptoms similar to those encountered with food-borne botulism.

The symptoms begin 18 to 24 hours after ingestion of the toxin, although several days may pass before symptoms occur. The initial symptoms include double vision, lack of coordination of eye muscles, inability to swallow, speech difficulty, generalized weakness, and dizziness, followed by descending progressive weakness of the extremities and weakness of the respiratory muscles. A patient may remain totally alert and oriented; there is no fever. Neurological examination will show flaccid muscle weakness of the tongue, larynx, respiratory muscles, and extremities. Patients remain fully conscious until shortly before death occurs from respiratory paralysis or cardiac arrest. The mortality rate can be high. Patients who recover do not develop productive antibodies (antitoxin) in the blood.[3]

Treatment — Since *C. botulinum* toxin blocks the actions of nerves that activate muscles necessary for breathing, an antitoxin can be injected up to about 24 hours (based on monkey studies) after exposure to a lethal toxin dose and still prevent death. The two types of available antitoxins prepared from horse sera are trivalent (includes types A, B, E) and heptavalent (types A, B, C, D, E, F, and G) preparations. It should be noted that patients face a theoretical risk of developing serum sickness from such antitoxins.

A vaccine preparation now available allows an individual to develop antibodies to the five most common *C. botulinum* types (A, B, C, D, and E). Studies have shown that the vaccine regimen produces protective antitoxin levels in greater than 90% of those vaccinated.[3]

3.3.3.2 *Clostridium perfringens* toxin

Disease caused	Toxin-mediated pulmonary syndrome
Variations of disease	N/A
Incubation period	6 hours to 1 day
Routes of exposure	Ingestion, inhalation
Person-to-person transmissibility	No
Vaccines available	No

This bacterium can produce numerous types of toxins that cause a range of medical problems. The most potent toxin of *C. perfringens* is the alpha toxin — the one most likely to be used by terrorists. It would be lethal by aerosol delivery.

Incubation period is 6 hours to 1 day. Symptoms of exposure include the development of gas gangrene (anaerobic destruction of tissue with the production of gas) and myonecrosis (muscle destruction). Naturally-occurring gas gangrene usually involves rapid invasion, liquefaction (changing solids to liquids), and necrosis of muscles, with gas formation followed by clinical signs of toxicity.

In an infected open wound scenario, the infection spreads in 1 to 3 days from an original contaminated wound site into the subcutaneous tissue and muscle. The result is a foul-smelling discharge progressing to necrosis and fever. Toxemia, shock, and death may follow. With a naturally-occurring food-borne illness, a patient usually develops a crampy abdominal syndrome within 6 to 18 hours after eating *C. perfringens*-laden food. The stools are foul smelling and the diarrhea has a foamy character. Vomiting and fever are found only occasionally. The toxin forms when the bacteria sporulate in the gut. The entire illness usually lasts 1 to 2 days.[3]

Treatment — There is no specific treatment for *C. perfringens* toxins, even though the organism is susceptible to penicillin, which is the drug of choice for a naturally acquired infection. Recent laboratory data indicate that clindamycin and rifampin may suppress toxin formation. Available polyvalent antitoxins contain antibodies to several toxins that have been used in treatment, but not enough data exist to prove efficacy.[3]

3.3.3.3 *Staphylococcus* Enterotoxin B/TSST-1

Disease caused	Toxin-mediated systemic disease
Variations of disease	Enterotoxins A, B, C, D, and E
Incubation period	1 to 6 hours
Routes of exposure	Ingestion, inhalation
Person-to-person transmissibility	No
Vaccines available	Yes (experimental)

Staphylococcus enterotoxin is one of the most toxin-producing germs known. It is resistant to stomach acids and can survive in boiling water for 30 minutes. If a terrorist dispensed *S.* enterotoxin B in aerosol form, the symptoms would be similar to those of food poisoning and toxic shock syndrome, but much more severe. After aerosol delivery of the toxin, a patient would experience a very rapid onset of symptoms 1 to 6 hours after exposure. The symptoms include fever, headache, myalgia, and nonproductive cough. A sunburn-like rash may appear and will desquamate (peel) after 10 to 14 days. The fever is generally less than 102°F, but may reach 106°F. The fever may last several days and depends on the amount of toxin inhaled. Patients will also likely experience nausea, vomiting, and diarrhea. In severe cases, they will experience fluid accumulation in the lungs and possibly acute respiratory distress syndrome that may lead to respiratory failure. If high levels of toxins are ingested, low blood pressure leading to septic shock and death can occur.[3]

Treatment — Because patients generally are exposed to small amounts of *S.* enterotoxin B toxin as a result of eating contaminated food (e.g., ham, salads with mayonnaise, and unrefrigerated cream dishes), they often have some protective antibodies. However, if the antibodies are overwhelmed by a terrorist-induced dosage, treatment with steroids should mollify patients' symptoms. Further treatment would be supportive care, especially compensating for oxygen and fluid losses. While no vaccines are available, experimental vaccines in development have shown promise.[3]

3.3.3.4 Ricin

Disease caused	Poisoning
Variations of disease	N/A
Incubation period	2 to 8 hours
Routes of exposure	Inhalation, ingestion, injection
Person-to-person transmissibility	No
Vaccines available	Only in experimental form

Ricin would probably be aerosolized, but it could also be used by an assassin who might prepare it for ingestion or injection. Ricin is not nearly as potent as the *C. botulinum* toxin, and thus would have to be produced in very large quantities for large-scale use.

About 2 to 8 hours after exposure, a patient would experience the abrupt onset of nausea and vomiting, abdominal cramps, diarrhea, chest tightness, vascular collapse, shortness of breath, and arthralgias, followed by a profuse sweating episode. If ricin is delivered by ingestion or injection, severe lung involvement would be absent. Ingestion causes gastrointestinal hemorrhage with hepatic, splenic, and renal necrosis. If ricin is injected, severe local necrosis of muscle and lymph nodes and organ damage occurs.[3]

Treatment — No vaccine or antitoxin is available; both are under development. Supportive management depends on the route of exposure and therefore is varied.[3]

3.3.3.5 Saxitoxins

Disease caused	Poisoning
Variations of disease	N/A
Incubation period	Minutes to 1 hour
Routes of exposure	Ingestion, inhalation
Person-to-person transmissibility	No
Vaccines available	No

Saxitoxins are water-soluble compounds that prevent proper nerve functioning. They are produced in nature by plant-like marine protozoa called dinoflagellates. Humans typically acquire such toxins by eating bivalve mollusks fed on dinoflagellates. A terrorist would likely deliver a saxitoxin as an aerosol or use it as a poison to contaminate food or water.

Symptoms of exposure occur within minutes to an hour after ingestion or inhalation, and depend on the quantity of toxin ingested or inhaled. The first symptoms include numbness or tingling of the lips, tongue, and fingertips. The numbness progresses to the extremities with or without a burning sensation. This is followed by muscular incoordination. Nausea and vomiting occur only in a minority of cases. Some individuals report floating sensations, generalized weakness, dizziness, inability to speak properly, memory loss, and headache. Patients remain conscious. Those who survive the first 12 to 24 hours usually recover, because the toxin is cleared quickly by the body.[3]

Treatment — As with other toxin therapies, treatment is limited mainly to supportive management. Mechanical respiration support may be required in severe cases. An antitoxin has shown success in animal models, but no human data are available to date. No vaccine has been developed. Induction of vomiting may prove to be useful as part of the treatment program. Ipecac syrup is a good emetic that could be used to induce vomiting.[3]

3.3.3.6 Trichothecene Mycotoxins (T-2)

Disease caused	Poisoning
Variations of disease	N/A
Incubation period	2 to 4 hours
Routes of exposure	Inhalation, ingestion
Person-to-person transmissibility	No
Vaccines available	No (experimental antitoxin)

Trichothecene mycotoxins are produced by a number of fungal molds of the *Fusarium*, *Myrotecium*, *Trichoderma*, and *Stachybotrys* genera. They inhibit protein synthesis, impair DNA synthesis, and interfere with cell membrane structures and functions. The potential routes of exposure are inhalation, ingestion, and skin absorption. A terrorist may take advantage of any of these routes.

Mycotoxins act very quickly on the skin, mucous membranes, and bone marrow. They are different from most toxins in that they can adhere to and penetrate the skin in addition to being inhaled or ingested. For this reason, mycotoxins should be treated more like chemicals than biological agents. Ingestion can result in nausea, vomiting, and diarrhea (sometimes bloody), with abdominal cramping. Mouth and throat pain is evident, and saliva or sputum can be blood-tinged. Eye pain, tearing, and blurred vision can also occur. Inhalation is usually characterized by difficulty in breathing, wheezing, and coughing. Nasal intake produces pain, sneezing, and nasal bleeding. Systemic toxicity can result if sufficient toxin is ingested or inhaled. Such toxicity manifests as generalized weakness, dizziness, overall loss of coordination, low blood pressure, rapid heartbeat, and low body temperature. Death can occur in minutes to days.[3]

Treatment — Oral superactivated charcoal preparation is standard therapy for poison ingestion. This type of preparation is five times stronger than ordinary charcoal. Supportive therapy must be provided, depending on the symptoms displayed. No antitoxin is currently available for human use. However, an experimental antitoxin has shown some promise in animals.[3]

TABLE 3.3
Symptoms of Exposure and Treatment Options

Contaminants	Routes of Exposure	Symptoms of Exposure	Treatment Options
		Chemical Agents	
Blister Agents			
Mustards			
Distilled mustard (HD)	Inhalation, skin absorption, ingestion, local skin and eye impacts	Vapor Contact: Severe choking Temporary/permanent blindness Inflammation of respiratory tract Stripping bronchial tubes of mucous membrane Liquid/Impregnated Solid Contact: Inflammation of tissue around eyes Reddening of skin Ulceration of skin into watery boils Blistering of throat and lungs Dry land drowning Destruction of white blood cells Bone marrow destruction Damage to immune system Often leads to death	No known antidotes Treatment consists of symptomatic management of lesions, controlling infection with antibiotics, and controlling pain with local anesthetics
Arsenicals			
Methyldichloro-arsine (MD)	Inhalation, skin absorption, ingestion, local skin and eye impacts	Vapor Contact: Severe respiratory pain Severe damage to lung membranes Severe eye discomfort Dry land drowning	British Anti-Lewisite dimercaprol alleviates some effects (Ellison 2000) Control infection with antibiotics Control pain with local anesthetics

		Liquid Contact: Skin blistering Permanent corneal damage in case of eye contact Bone calcium displacement with arsenic Bone marrow destruction May lead to death	
Phenyldichloro-arsine (PD)	Inhalation, skin absorption, ingestion, local skin and eye impacts	Skin blistering Damage to eyes (cornea), nasal membranes, throat, and lungs Vomiting Bone calcium displacement with arsenic Bone marrow destruction May lead to death; not as lethal as HD	British Anti-Lewisite dimercaprol alleviates some effects (Ellison 2000) Control infection with antibiotics Control pain with local anesthetics
Ethyldichloro-arsine (ED)	Inhalation, skin absorption, ingestion, local skin and eye impacts	Severe discomfort to eyes Permanent corneal damage Harsh respiratory effects Damage to bone marrow Damage to digestive tract and endocrine systems Redness of skin followed by blistering Lung tissue destruction Dry land drowning	British Anti-Lewisite dimercaprol alleviates some effects (Ellison 2000) Control infection with antibiotics Control pain with local anesthetics
Lewisite (L)	Inhalation, skin absorption, ingestion, local skin and eye impacts	Immediate pain upon skin contact Skin rash followed by blistering Chemical burns on skin Blindness Immediate destruction of lung tissue Systemic blood poisoning	British Anti-Lewisite dimercaprol alleviates some effects (Ellison 2000) Control infection with antibiotics Control pain with local anesthetics

-- continued

TABLE 3.3 (continued)
Symptoms of Exposure and Treatment Options

Contaminants	Routes of Exposure	Symptoms of Exposure	Treatment Options
		Chemical Agents	
Lewisite (L)		Pulmonary edema Neural disorder Subnormal body temperature Low blood pressure Dry land drowning	
Nitrogen Mustards			
2,2-dichlorotriethylamine (HN-1)	Inhalation, skin absorption, ingestion, local skin and eye impacts	Cumulative poison Highly irritating to eyes and throat Skin rash followed by blistering Degradation of blood oxidase function Lesions in small intestine Degenerative changes to mucous membranes Severe bloody diarrhea Bronchial pneumonia Dry land drowning Death of living tissue	No known antidotes Control infection with antibiotics Control pain with local anesthetics
2,2-dichloro-N-methyldiethylamine (HN-2)	Inhalation, skin absorption, ingestion, local skin and eye impacts	Cumulative poison Highly irritating to eyes and throat Blindness Skin rash followed by blistering Interferes with blood hemoglobin functioning Dry land drowning Asphyxiation Heart failure	No known antidotes Control infection with antibiotics Control pain with local anesthetics

2,2,2-trichlorotriethylamine (HN-3)	Inhalation, skin absorption, ingestion, local skin and eye impacts	Cumulative poison Highly irritating to eyes and throat Permanent corneal damage Skin rash followed by blistering Interferes with blood hemoglobin functioning Destroys white blood cells, causing damage to immune system Dry land drowning Asphyxiation Heart failure	No known antidotes Control infection with antibiotics Control pain with local anesthetics
Oximes			
Phosgene oxime (CX)	Inhalation, ingestion, local skin and eye impacts	Affects skin, muscle, and nerves Severe pain on contact Bleached appearance of skin followed by red rash ring that turns dark, then forms scab that eventually falls off Immediate death from systemic shock or trauma	No known antidotes Control infection with antibiotics Control pain with local anesthetics
Blood Agents			
Hydrogen cyanide (AC)	Inhalation, skin absorption, ingestion	Irritating to eyes and upper respiratory tract Weakness Headache Disorientation Nausea Vomiting Affects ability to utilize oxygen Loss of consciousness Shock Asphyxiation Death	The Lilly Cyanide Antidote Kit contains amyl nitrite, sodium nitrite, and sodium thiosulfate Dimethylaminophenol, cobalt-edetate, and vitamin B_{12a} are alternative antidotes (Ellison 2000)

-- continued

TABLE 3.3 (continued)
Symptoms of Exposure and Treatment Options

Contaminants	Routes of Exposure	Symptoms of Exposure	Treatment Options
		Chemical Agents	
Cyanogen chloride (CK)	Inhalation, skin absorption, ingestion, local skin and eye impacts	Irritating to eyes and upper respiratory tract Flushed skin Weakness Headache Disorientation Nausea Vomiting Affects ability to utilize oxygen Loss of consciousness Shock Asphyxiation Death	The Lilly Cyanide Antidote Kit contains amyl nitrite, sodium nitrite, and sodium thiosulfate Dimethylaminophenol, cobalt-edetate, or vitamin B_{12a} are alternative antidotes (Ellison 2000).
Arsine (SA)	Inhalation	Cumulative poison Flushed skin Headache Irritability Chills Nausea Vomiting Displaces calcium in bone matter Impacts production of new blood cells Damages liver and kidneys Shock Asphyxiation Death	No known antidotes

Choking Agents			
Chlorine	Inhalation, local skin and eye impacts	Vomiting Diarrhea Strips lining from bronchial tubes and lungs Immediate inflammation of bronchial tubes and lungs Massive amounts of phlegm Hemorrhage Asphyxiation Dry land drowning	No known antidotes
Phosgene (CG)	Inhalation	Cumulative poison Mild irritation to eyes Attacks lung capillaries Lungs fill with watery fluid Sudden death	No known antidotes
Diphosgene (DP)	Inhalation	Cumulative poison Mild irritation to eyes Attacks lung capillaries Lungs fill with watery fluid Sudden death	No known antidotes
Nerve Agents			
Tabun (GA)	Inhalation, skin absorption (liquid/vapor), ingestion	Interference with neural synapses Over-stimulation of nervous system Malfunctioning of various organs Massive congestion of enzymes and fluids in all major organs of nervous system, particularly brain Death	Initial treatment is antidote consisting of atropine and 2-PAM chloride

-- continued

TABLE 3.3 (continued)
Symptoms of Exposure and Treatment Options

Contaminants	Routes of Exposure	Symptoms of Exposure	Treatment Options
		Chemical Agents	
Sarin (GB)	Inhalation, skin absorption (liquid/vapor), ingestion	Cumulative agent Interference with neural synapses Over-stimulation of nervous system Malfunctioning of various body organs Massive congestion of body enzymes and fluids in all major organs of nervous system, particularly brain Death	Initial treatment is antidote consisting of atropine and 2-PAM chloride
Soman (GD)	Inhalation, skin absorption (liquid/vapor), ingestion	Cumulative agent; more neurologically active than GA or GB Interference with neural synapses Over-stimulation of nervous system Spasmodic symptoms such as jerking, urinating, and defecation Malfunctioning of various body organs Massive congestion of body enzymes and fluids in all major organs of nervous system, particularly brain Death	Initial treatment is antidote consisting of atropine and 2-PAM chloride
V gas (VX)	Inhalation, skin absorption (liquid/vapor), ingestion	Cumulative agent Spasmodic symptoms such as jerking, urinating, and defecation Heart, lungs, and brain can cease functioning Death	Initial treatment is antidote consisting of atropine and 2-PAM chloride
		Biological Agents	

Bacillus anthracis (anthrax)	Inhalation, ingestion, direct contact via break in skin, fly bite	<u>Inhaled anthrax:</u> Incubation period is 1–5 days Symptoms for initial phase: Muscle pain Low-grade fever Nonproductive cough Mild chest pain Symptoms for acute phase: Difficulty breathing Profuse sweating Turning blue High fever Increased pulse and respiratory rate Shock and death <u>Skin anthrax (break in skin):</u> Incubation period 2–7 days Symptoms for Initial phase: Itchy pimples on face, neck, or arms Break in skin blisters Blisters break to form open ulceration Ulceration develops black scab Swelling Fever Swelling of lymph nodes Death is rare <u>Gastrointestinal anthrax:</u> Incubation period 2–7 days Caused by ingestion of contaminated meat from infected animal Intestinal symptoms	Quinolene antibiotics: ciprofloxacin, levofloxacin, ofloxacin Tetracycline antibiotics: doxycycline Penicillin antibiotics: amoxicillin, penicillin V, Penicillin G Vaccines are available; six doses at 0, 2, and 4 weeks, then 6, 12, and 18 months, followed by annual boosters See Tierno 2002 or other medical references for details on administration of medications and/or vaccines

-- continued

TABLE 3.3 (continued)
Symptoms of Exposure and Treatment Options

Contaminants	Routes of Exposure	Symptoms of Exposure	Treatment Options
		Biological Agents	
Bacillus anthracis (anthrax)		Nausea Vomiting Loss of appetite Fever Abdominal pain Bloody diarrhea/vomiting blood Fluid build-up in abdominal cavity Shock Toxemia Death (25– 60%) Oropharyngeal symptoms Swelling of neck Lesion in oral cavity Fever Swollen lymph nodes Inability to swallow Shock Toxemia Death (25– 60%)	
Yersinia pestis (bubonic plague)	Inhalation, bite of infected flea	Incubation period 2–10 days Malaise High fever Tender compressible lymph nodes (groin, neck, underarm) Vomiting Diarrhea	One or more antibiotics: tetracycline, streptomycin, gentamicin, chloramphenicol, quinolone A vaccine seems to be effective SEe Tierno 2002 or other medical references for details on administering medications and/or vaccines

		Shock Renal failure Heart failure Death	
Yersinia pestis (pneumonic plague)	Inhalation, bite of infected flea	Incubation period 1– 3 days High fever Cough Chest pain Headache Bloody sputum Chills Lack of energy Myalgia Pneumonia Difficulty breathing Turning blue Death (100% when untreated)	
Brucella melitensis (brucellosis)	Ingestion, direct contact via break in skin, mucous membranes	Incubation period 1– 6 weeks Fever Chills Sweating Headache Fatigue Myalgia Arthralgia Swollen lymph, spleen, and liver Death (about 6%) if *B. melitensis* is involved; less than 1% for other species; most deaths associated with infection of heart lining or infection of membranes around the brain	Often treated with doxycycline, rifampin, and ofloxacin See Tierno 2002 or other medical references for details on administering medications

-- continued

TABLE 3.3 (continued)
Symptoms of Exposure and Treatment Options

Contaminants	Routes of Exposure	Symptoms of Exposure	Treatment Options
		Biological Agents	
Francisella tularensis (tularemia)	Inhalation, ingestion, direct contact via break in skin, bites of infected ticks, mosquitoes, or flies	Swollen lymph nodes Fever (usually low) Malaise Headache Pain involving regional lymph nodes Pneumonia	Often treated with doxycycline and tetracycline See Tierno 2002 or other medical references for details on administering medications
Coxiella burnetii (Q fever)	Inhalation, ingestion	Incubation period 2–14 days Fever Cough Chills Myalgia Headache Chest pain Does not generally lead to critical illness	Antibiotic treatments include tetracycline, doxycycline, erythromycin, rifampin, trimethoprin–sulfamethoxazole See Tierno 2002 or other medical references for details on administering medications
Vibrio cholerae (cholera)	Ingestion	Incubation period 4 hours– 5 days Symptoms of exposure: Nausea Vomiting Profuse diarrhea without abdominal cramps; stool resembles "ricewater" with much mucus Rapid loss of fluid and electrolytes Dehydration leads to circulatory collapse and kidney shutdown Mortality rate without treatment is as high as 50%	Therapy should include replacement of fluids and electrolytes and antibiotics such as tetracycline, doxycycline, ciprofloxacin, and erythromycin See Tierno 2002 or other medical references for details on administering medications

Burkholdera mallei (glanders)	Skin abrasion, inhalation	Incubation period 3– 14 days Acute form: Infection of nasal, oral, or conjunctival mucous membranes Blood-streaked discharge from the nose, nodules, and ulcerations Chronic form: Affects joints and lymphas Ulcerated skin nodules with pus Fever Sweats Myalgia Headache Enlarged spleen Chest pain Sometimes pneumonia Almost 100% fatal if untreated	Sulfadiazine, doxycycline, rifampin, trimethoprim–sulfamethoxazole, streptomycin, ciprofloxacin See Tierno 2002 or other medical references for details on administering medications
Burkholdera pseudomallei (melioidosis)	Ingestion, inhalation, fly bite	Incubation period 3– 14 days Pneumonia Bloodstream infection Hypotension Shock Mortality rate for untreated acute disease patients is 95%	Tetracycline, chloramphenicol, trimethoprim–sulfamethoxazole, doxycycline, ceftazidime See Tierno 2002 or other medical references for details on administering medications
		Viral Agents	
Variola major (smallpox)	Inhalation, contact with skin lesions or secretions	Incubation period 10–14 days Malaise Fever Headache Chills Backache	Vaccinia immune globulin, vaccinia vaccine See Tierno 2002 or other medical references for details on administering medications and/or vaccines

-- continued

TABLE 3.3 (continued)
Symptoms of Exposure and Treatment Options

Contaminants	Routes of Exposure	Symptoms of Exposure	Treatment Options
		Viral Agents	
Variola major (smallpox)		Eruption of skin or rash (begins with pimples that turn to blisters that fill with pus and form crust that falls off, leaving scar) Death (15–40%, unvaccinated patients and <1%, vaccinated patients)	
Venezuelan equine encephalitis	Inhalation, mosquito bite	Incubation period 4–21 days Rapid onset of fever (usually high) Headache Dizziness Lethargy Depression Anorexia Chills Myalgia Photophobia Nausea Vomiting Cough Sore throat Diarrhea Convulsions <5% of cases showed convulsions, coma, and paralysis Low death rate	There is no specific antiviral treatment One can only treat the symptoms of headache or myalgia, controlling convulsions, aiding difficulty in breathing

Crimean Congo hemorrhagic fever	Inhalation, tick bite	Incubation period 4–21 days Fever Myalgia Low blood pressure Flushing Black-and-blue marks Gastrointestinal bleeding Extensive ecchymoses Headache Back pain Nausea Vomiting Delirium Jaundiced appearance Mortality rate is 15–30%; for some (e.g., Ebola), can be as high as 90%	Immediate attention should be given to the management of hypotension caused by fluid loss Few drugs are effective Ribavirin can be given to Crimean Congo hemorrhagic fever patients Treatments with immune globulin vaccines are useful against Crimean Congo hemorrhagic fever, Rift Valley fever, Bolivian hemorrhagic fever, and Lassa fever Yellow fever vaccine is the only established and licensed vaccine for a hemorrhagic fever; several others are under development
Rift Valley hemorrhagic fever	Inhalation, mosquito bite	Incubation period 4–21 days High fever Hemorrhagic areas under skin Fewer than 1% develop more serious viral hemorrhagic fever syndrome leading to mucosal bleeding, liver and kidney failure, and shock before death	Intravenous ribavirin An effective inactivated vaccine is available
		Toxins	
Clostridium botulinum	Inhalation, ingestion	Incubation period is 18–72 hours Double vision Lack of coordination of eye muscles Inability to swallow Speech difficulty Generalized weakness	Trivalent and heptavalent antitoxins are available A vaccine allows development of antibodies to the most common forms of *Clostridium botulinum* (Types A, B, C, D, and E) See Tierno 2002 or other medical references for details on administering medications

-- continued

TABLE 3.3 (continued)
Symptoms of Exposure and Treatment Options

Contaminants	Routes of Exposure	Symptoms of Exposure	Treatment Options
		Toxins	
Clostridium botulinum		Dizziness Weakness of the tongue, larynx, respiratory muscles, and extremities Death from respiratory paralysis or cardiac arrest (high mortality rate)	
Clostridium perfringens	Ingestion, inhalation	Incubation period 6 hours–1 day Symptoms of natural ingestion exposure from contaminated food: Gas gangrene (gas produced by anaerobic destruction of tissue) Myonecrosis (muscle destruction) Open wounds with foul-smelling discharges Fever Stomach gas formation Crampy abdominal syndrome Foul smelling foamy diarrhea Vomiting or fever (not common) Toxemia Shock Death	Penicillin is the drug of choice for naturally acquired infection Studies show that clindamycin and rifampin may suppress toxin formation
Staphylococcus enterotoxin B	Ingestion, inhalation	Incubation period 1–6 hours Fever Headache Myalgia Nonproductive cough Sunburn-like rash Desquamation (peeling skin)	Treat with steroids to mollify symptoms, along with supportive care to compensate for oxygen and fluid loss Vaccines are in the development phase

		Nausea Vomiting Diarrhea Fluid accumulation in lungs Acute respiratory distress Respiratory failure Low blood pressure Septic shock Death	
Ricin	Inhalation, ingestion, injection	Incubation period 2–8 hours Nausea Vomiting Abdominal cramps Diarrhea Chest tightness Vascular collapse Shortness of breath Arthralgia Profuse sweating Gastrointestinal hemorrhage Hepatic, splenic, and renal necrosis Organ damage	Vaccine and antitoxin are under development Supportive management varies depending on the route of exposure
Saxitoxins	Ingestion, inhalation	Incubation period minutes – 1 hour Numbness or tingling of lips, tongue, and fingertips; progresses to extremities with or without burning sensation Muscular incoordination Nausea and vomiting (not common) Floating sensation Generalized weakness	Treatment is limited mainly to supportive management Mechanical respiration support may be required in severe cases No vaccine has been developed Induction of vomiting may prove to useful Ipecac syrup can induce vomiting

-- continued

TABLE 3.3 (continued)
Symptoms of Exposure and Treatment Options

Contaminants	Routes of Exposure	Symptoms of Exposure	Treatment Options
		Toxins	
Saxitoxins		Dizziness Inability to speak properly Memory loss Headache	
Trichothecene mycotoxins (T-2)	Inhalation, ingestion	Incubation period 2– 4 hours <u>Ingestion exposure</u>: Nausea Vomiting Diarrhea (sometimes bloody) Abdominal cramping Mouth and throat pain Blood-tinged saliva or sputum Eye pain, tearing, and blurred vision <u>Inhalation exposure:</u> Difficulty breathing Wheezing Coughing <u>Nasal intake:</u> Pain and sneezing, nasal bleeding Systemic toxicity if sufficient toxin is ingested or inhaled; characterized by weakness, dizziness, loss of coordination, low blood pressure, rapid heartbeat, and low body temperature; death can occur in minutes to days	

REFERENCES

1. Compton, J.A.F., *Military Chemical and Biological Agents*, Telford Press, Caldwell, NJ, 1987.
2. Ellison, D.H., *Handbook of Chemical and Biological Warfare Agents*, CRC Press, Boca Raton, FL, 2000.
3. Tierno, P.M., Jr., *Protect Yourself against Bioterrorism*, Pocket Books, New York, 2002.
4. NCRP, *Exposure of the Population in the United States and Canada from Natural Background Radiation*, Report 94, National Council on Radiation Protection and Measurements, Bethesda, MD, 1987.
5. ICRP, *1990 Recommendations of the International Commission on Radiological Protection*, Publication 60, Pergamon Press, New York, 1991.
6. AFRRI, *Medical Management of Radiological Casualties Handbook*, Special Publication 99-2, Armed Forces Radiobiology Research Laboratory, Bethesda, MD, 1999.
7. Turner, J.E., *Atoms, Radiation, and Radiation Protection*, McGraw-Hill, New York, 1992.

4 Minimizing Exposure to Radiation and Warfare Agents

This chapter provides guidance on how emergency responders can best minimize exposure to weapons of mass destruction that utilize no explosives (e.g., aerosol delivery of chemical or biological agent) or include conventional explosives (e.g., dirty bombs). Chapter 5 provides guidance on minimizing exposure when responding to a nuclear explosion. Three factors are inherent in radiation protection philosophy are utilized to keep exposures as low as reasonably achievable (ALARA): *time*, *distance*, and *shielding*. While this philosophy was derived from the radiation protection industry, it is also effective in minimizing exposure to chemical and biological agents.

The *time* factor is the duration of exposure, with the assumption that the shorter the exposure time, the less likely it is that an individual will suffer an ill effect. The *distance* factor is the physical separation (e.g., in feet or miles) between an individual and the location of the terrorist event. The *shielding* factor refers to physical barriers between an individual and hazardous substances emanating from the attack location. These three factors are discussed in detail below, along with general rules for minimizing exposure.

4.1 TIME OF EXPOSURE

Time in this context is duration to which an individual is exposed to a hazardous substance. Time is relevant because the longer an individual remains in the open or at the site of a terrorist attack, the more likely there will be an exposure to a hazardous substance. Time is also relevant because the longer an individual is in contact with a hazardous substance (e.g., on the skin), the larger the exposure. The time factor is applied whether an individual is directly at the site of the attack or only in the vicinity. However, the time factor is applied differently in these two cases. For example, if an individual is in the vicinity of the attack site, the following time factor rule applies:

Time Rule 1: Stay inside or move inside as quickly as possible and stay there until authorities confirm it is safe to evacuate.

If an explosion or other incident releases hazardous substances into the air, the most likely route of exposure is through inhalation of airborne contaminants. If the air is contaminated, the best place to be is in a building, preferably with the heating and air conditioning system set to recirculate the existing air in the building to minimize the amount of contaminated air pulled into the building. The other option is to turn the heating and air conditioning system off. This simple guidance is the same used, for example, by someone with allergies when pollen levels are high. The best practice is to stay indoors until authorities confirm it is safe to evacuate the area if required. The authorities may be emergency responders at the site or public service announcers on the radio or television.

Different rules apply to individuals at the site of an attack. They must be concerned about potential radiation exposure, inhalation of airborne contaminants, contamination on skin and clothing, and incidental ingestion of contaminated substances. For individuals at the site of an attack, the following time factor rules apply:

Time Rule 2: Leave the damaged building or affected area in a quick and orderly manner and seek shelter in a nearby, preferably undamaged, building (following an emergency response plan if one exists).

Time Rule 3: Minimize the time of exposure by removing soiled articles of clothing and washing all exposed body parts, including the mouth and hair, as soon as possible.

Moving away from the site of the attack as quickly as possible will reduce the amount of exposure. Orderly retreat will reduce the risk of other accidents such as sprains or broken bones. An individual could receive additional exposure for every minute spent in the open near the attack site. If he has had direct contact with a hazardous substance, the time of exposure should be minimized. Hazardous substances on the skin, clothing, and hair should be removed as quickly as possible using specific decontamination methods outlined in Chapter 7. Special medical attention should be administered as soon as possible if an individual has ingested or inhaled a hazardous substance or if the hazardous substance entered the body through a wound. Consider the following example:

> A man works in an office building where a dirty bomb explodes. The bomb does not directly damage his office, so he decides to call his wife to tell her what happened and inform her that he is unhurt. In the 3 minutes that it takes for him to complete the call and gather his belongings, smoke from the bomb has reached his office and the staircase he must use to exit the building. It takes him 2 minutes to exit the building. During all that time he is exposed to contaminated smoke. Radioactive-contaminated dust is now in his hair, on his skin and clothes, in his mouth, and in his lungs. After he exits the building, he seeks out coworkers and talks about the incident while watching emergency responders near the blast site. After an hour, he asks an emergency worker how things are going and is told that he has been exposed to radioactive materials. He breathed contaminated air for 2 minutes, stood around covered with contaminated dust for 1 hour, and may have been exposed to radiation emanating from the blast site for 1 hour.

The man in this example would have received minimal exposure if he had left the building immediately and evacuated to a nearby building or to a safe distance.

He could have called his wife after evacuating the area, which would have only been a few minutes later than calling her from his office. He also should have called 911 for emergency assistance and, if soiled, removed his contaminated clothing and washed his skin and hair as quickly as possible. The man in this example failed the test of time. The effective use of available time can mean no exposure instead of limited or extensive exposure.

4.2 DISTANCE

Distance in this context is the physical separation (e.g., in feet or miles) between an individual and the location of a terrorist attack. Distance is relevant because the closer an individual is to the site of the attack, the more likely is exposure to a hazardous substance. Consider two rules for the distance factor:

Distance Rule 1: Do not remain near the site of the attack — quickly put distance between yourself and any potentially hazardous substance, including smoke and debris.

Distance Rule 2: Direct contact is not required to receive radiation exposure — increased distance equates to decreased exposure to radiation.

Hazardous substances can be distributed in plumes of dust or smoke, or can be contained in or on debris scattered about the site of an attack. An individual who uses common sense when he sees a plume of smoke should try to avoid direct contact by retreating into a nearby building. Because fine particles may also be in the air and not easily observed, the potential for exposure reduces as distance from the site increases. Risk from exposure may also be reduced by seeking shelter in a nearby building.

Everyone should avoid plumes of smoke that may deposit contaminants at larger distances. It would be unfortunate for an individual to safely evacuate from a terrorist attack only to be caught in the open a mile from the attack and exposed to a plume of contaminated smoke. As with the time factor, the best practice is to move quickly from the site of the attack and seek shelter in a building with the heating and air conditioning system either set to recirculate the existing air in the building or turned off. The next step is to wait for authorities to confirm when it is safe to evacuate the area.

Radiation from a dirty bomb can emanate from a blast site in a contaminated plume of smoke or in contaminated debris. Radiation cannot be detected without special instruments, and radiation exposures can occur even without direct contact. Therefore, leaving a damaged building does not eliminate the risk of exposure. An effective tool to minimize or eliminate the potential for hazardous substance exposure is to move away from the site of the attack and into a building that provides protection from airborne contaminants.

Consider again the man in the previous example of the office bombing. He should have left the site of the explosion immediately and entered a nearby building with the heating and air conditioning system either set to recirculate the existing air in the building or turned off. Once inside, he would have been protected from contaminated smoke and debris and then should have called 911 for emergency assistance. He could have begun removing soiled articles of clothing or contamination on his skin. Because he remained in the area for an hour, he could have been exposed to radiation emanating from the site or from contaminated smoke, or contaminated debris.

Placing distance between yourself and potentially contaminated materials can significantly reduce or eliminate exposure to hazardous substances.

4.3 SHIELDING

Shielding in this context refers to placement of a physical barrier between an individual and hazardous substances emanating from contaminated materials. In the field of radiation protection, shielding typically refers to a barrier that reduces radiation levels. The term is used in a broader sense here to refer to any physical barrier that reduces or eliminates exposure to a hazardous substance. Consider three rules of thumb for the shielding factor:

Shielding Rule 1: To shield against a radiation or biological attack, move to the dark corners of a building's basement (if available) or to a windowless center room. For attacks involving chemical agents, seek shelter in a windowless center room on the ground floor — not in a basement where chemical agents will concentrate.

Shielding Rule 2: Heat the air of a sealed building to create positive pressure and prevent the infiltration of contaminants. Always use recirculated air or air purified by a high-efficiency particle arrestor (HEPA) filter. If this is not possible, turn the heating and air conditioning system off.

Shielding Rule 3: Use available resources to shield your lungs against airborne contaminants (e.g., cover your mouth with a handkerchief) and shield your body from radiation (e.g., move behind a concrete wall).

A shield can be a building, a wall, a folded handkerchief, etc. The shield can serve as protection against radiation, a chemical or biological agent, smoke, dirt, or even flying debris. The concept is simply to place a barrier between you and a potential hazard. The effectiveness of a shielding material depends on the form of the hazardous substance. A shield against falling debris is a durable overhead cover. The inside of a nearby building may be a suitable shield against an explosion or the resulting shower of debris.

A building may also be an effective shield against a cloud of contaminated dust or smoke. Airtight doors and a heating and air conditioning system set to recirculate the existing air in a building can significantly reduce potential exposure to airborne contaminants. The walls of a building will also provide shielding from radiation emanating from the site of the attack or from other contaminated materials. To be most secure from radiation, an individual should move away from exterior windows and doors and into the basement or center portion of a building (lateral, not vertical center). If the building has a basement, the surrounding earth will act as an effective radiation shield. Individuals should seek out the dark corners and hide behind solid material to provide the best shielding against radiation.

If a terrorist attack involves one or more chemical agents (e.g., phosgene, lewisite, distilled mustard gas) that have vapor densities much heavier than air, seek shelter in the center portion of the building, on the ground floor as opposed to the basement or other substructure where agents entering the building would concentrate. The basement or the central portion of the building on the ground floor is acceptable shelter for a biological agent (e.g., anthrax, plague) attack.

A small residence can even act as a shield. An individual can turn off the air circulation system (or switch it to recirculation mode), and close vents, windows, and doors to create a barrier that will help prevent contaminated smoke or other airborne debris from entering the home. Because a building may not always be nearby in the event of a terrorist attack, shielding could be provided from another object such as a car (with the vents closed), a folded handkerchief over the nose and mouth, or any object that reduces the potential of exposure. The shield does not have to be sophisticated; it only needs to be effective.

Again consider the previous example of the man in the office bombing. While exiting the smoke-filled building, he could have covered his nose and mouth with a towel from his gym bag or a folded handkerchief. Upon exiting, he could have sought shelter in a nearby building. This would have shielded him against airborne contaminants and radiation emanating from contaminated smoke and debris.

If properly used, a very simple shield can significantly reduce or eliminate exposure to hazardous substances.

4.4 SUMMARY

Three factors inherent in radiation protection philosophy (*time*, *distance*, and *shielding*) can be used to protect against exposure to hazardous substances released during a terrorist attack. The *time* factor refers to the duration of exposure. The *distance* factor refers to the physical separation (in feet, miles, etc.) between an individual and contaminated materials. The *shielding* factor refers to physical barriers between an individual and hazardous substances emanating from the attack location.

The effective use of *time* can mean the difference between no exposure and extensive exposure. If an attack occurs in a nearby building or area, Time Rule 1 applies:

Time Rule 1: Stay inside or move inside as quickly as possible and stay there until authorities confirm it is safe to evacuate.

If an individual is at the site of an attack, Time Rules 2 and 3 apply:

Time Rule 2: Leave the damaged building or affected area in a quick and orderly manner and seek shelter in a nearby, preferably undamaged, building (following an emergency response plan if one exists).

Time Rule 3: Minimize the time of exposure by removing soiled articles of clothing, and washing all exposed body parts, including the mouth and hair, as soon as possible.

Placing distance between yourself and potentially contaminated materials can significantly reduce or eliminate exposure to hazardous substances.

Distance Rule 1: Do not remain near the site of the attack — quickly put distance between yourself and any potentially hazardous substances, including smoke and debris.

Distance Rule 2: Direct contact is not required to receive radiation exposure — increased distance equates to decreased exposure to radiation.

If properly used, a very simple shield can significantly reduce or eliminate exposures to hazardous substances.

Shielding Rule 1: To shield against a radiation or biological attack, move to the dark corners of a basement (if available) or to a windowless center room. For attacks involving chemical agents, seek shelter in a windowless center room on the ground floor as opposed to a basement or other substructure where chemical agents will concentrate.

Shielding Rule 2: Heat the air of a sealed building to create positive pressure and prevent the infiltration of contaminants, and always use recirculated air or air purified by a HEPA filter. If this is not possible, turn the heating and air conditioning system off.

Shielding Rule 3: Use available resources to shield your lungs against airborne contaminants (e.g., cover your mouth with a handkerchief) and shield your body from radiation (e.g., move behind a concrete wall).

5 Responding to a Nuclear Explosion

A worst-case terrorist attack would involve the use of a nuclear weapon. While guidance for attacks using conventional weapons generally applies to a nuclear attack, additional discussion is necessary because of the sheer magnitude of the devastation that would occur. The purpose of this chapter is to provide background information on nuclear weapons and practical guidance for responding to a nuclear attack.

The information in this chapter applies to everyone — average citizens, emergency responders, and soldiers. It is understood that emergency responders and military personnel may make sacrifices to save the lives of others, but these actions are taken on an individual basis in extremely dangerous circumstances. Providing for one's own safety achieves an increased likelihood of survival, a decrease in exposure and corresponding decrease in the likelihood of long-term health effects, and lower risk to others through diminished needs for help by emergency responders. It also allows emergency responders to have more time available to save lives by avoiding the most hazardous short-term exposures.

The response to a nuclear explosion differs dramatically from the response to an attack with conventional explosives. When conventional explosives such as dirty bombs disperse radiological materials, the health hazards from the radiation exposure are secondary to the explosion (flying shrapnel, debris, fire, and smoke). In the case of a nuclear explosion, the risks of death, serious short-term health effects, and serious long-term health effects are no longer secondary to the explosion.

5.1 NUCLEAR EXPLOSION BASICS

This section describes what happens to the local environment after the detonation of a nuclear weapon. Because it is only a summary, readers are encouraged to seek detailed information from libraries and the Internet on issues such as the workings of nuclear weapons and the technical aspects of radiation exposure. The following discussion is based on *Management of Terrorist Events Involving Radioactive Materials*, NRCP Report 138[1] and *Radioactive Hazards in Survival Planning.*[2]

5.1.1 General Discussion

Table 5.1 summarizes some of the variables associated with a nuclear explosion. A range of the yields (i.e., kilotons of TNT equivalent) is provided in the far left

TABLE 5.1
Variables Associated with Nuclear Explosion

Yield (kilotons)	Radius of Complete Destruction[a] (mi)	Radius of 50% Mortality from Air Blast[b] (mi)	Radius of 50% Mortality from Thermal Burns[b] (mi)	Radius of 400 rem from Initial Radiation[b] (mi)	Radius of 400 rem from Fallout in First Hour[b] (mi)
0.01	0.10	0.037	0.037	0.16	0.79
0.1	0.12	0.081	0.12	0.29	1.7
1.0	0.16	0.17	0.38	0.49	3.4
10	0.25	0.37	1.1	0.75	6.0
20	0.30	0.46	1.5	0.93	6.9
1000	1.0	1.6	10	1.3	13
10,000	2.5	3.4	29	1.5	17

Note: Actual fallout ranges depend on local meteorological conditions. Fallout direction relative to ground zero can be highly variable.

[a]Values for yields of 0.01 to 10 extrapolated from Brodsky, A. Radioactive hazards in survival planning, *Radiation Protection Management*, 18, 34, 2001. With permission.
[b]Values for yields of 20 to 10,000 extrapolated from NCRP, *Management of Terrorist Events Involving Radioactive Materials*, Report 138, 2001, National Council on Radiation Protection and Measurement. With permission.

column, and the other columns show ranges of potential threats based on distance from ground zero. The five primary threats from a nuclear explosion are:

- Destructive power of the explosion
- Air blast
- Thermal burns from intense heat
- Initial radiation from the nuclear reaction
- Radiation from fallout

The primary threats presented in Table 5.1 are briefly described, followed by a discussion of the effects from all threats combined.

5.1.1.1 Radius of Complete Destruction

The second column of Table 5.1 presents the radius of complete destruction following a nuclear explosion. It is assumed that an individual within this radius will not survive. If an individual near ground zero lives through the massive destructive power of the nuclear explosion, he will likely succumb to the intense heat and radiation. Table 5.2 supplements Table 5.1 and lists various levels of destruction,

TABLE 5.2
Level of Destruction versus Distance from Ground Zero

Yield (kilotons)	Radius of Complete Destruction[a] (mi)	Radius of Severe Destruction[a] (mi)	Radius of Moderate Destruction[a] (mi)	Radius of Light Destruction[a] (mi)
0.01	0.10	0.20	0.30	0.40
0.1	0.12	0.24	0.36	0.48
1.0	0.16	0.32	0.48	0.64
10	0.25	0.50	0.75	1.0
20	0.30	0.60	0.90	1.2
1000	1.0	2.0	3.0	4.0
10,000	2.5	5.0	7.5	10

[a]Based on values listed in Table 5.1 and relationships developed in Brodsky, A. Radioactive hazards in survival planning, *Radiation Protection Management*, 18, 34, 2001. With permission.

ranging from light to complete. The area within the outer rings is where protection against the air blast, thermal burns, initial radiation, and fallout is most important.

5.1.1.2 Air Blast

The third column of Table 5.1 presents the radius of 50% mortality from the air blast of a nuclear weapon. The blast (or shock wave) travels in all directions from ground zero at approximately the speed of sound. The blast alone can be fatal, but the risk of serious injury or fatality increases significantly because the shock wave picks up any materials in its path (e.g., shards of glass, metal, etc.). At the speed of sound, the air blast plus flying shrapnel travels at a speed of about 1 mile every 5 seconds. Thus, an individual 5 miles from ground zero has about 25 seconds to take cover before the air blast arrives. The same air blast speed applies to almost any size nuclear weapon.[2] The risk of injury from the blast drops with increased distance from ground zero.

5.1.1.3 Thermal Burns

The fourth column presents the radius of 50% mortality by thermal burns. The fireball from a nuclear explosion can reach temperatures in the tens of millions of degrees Fahrenheit and cause thermal burns at large distances. This intense heat can also cause temporary or permanent blindness and can ignite materials far from ground zero. Heat from the fireball will be felt instantly in all directions from ground zero; thus, the longer a person remains out in the open, the more intense the thermal burns will be. However, the heat from the fireball lasts only several seconds and can be shielded by solid materials like brick and earth (e.g., behind a wall or hill, in a ditch or subway tunnel, etc.). The risk of thermal burns drops with increased distance from ground zero.

5.1.1.4 Initial Radiation

The fifth column presents the radius of 50% mortality from the intense initial radiation present during the first few minutes following the explosion. As with thermal burns, radiation exposures begin instantly and occur in all directions from ground zero. Finding shelter quickly is critical. Initial radiation levels drop with increased distance from ground zero and can be shielded with the same materials that shield against thermal burns. The risk from exposure to initial radiation drops with increased distance from ground zero. After the initial radiation subsides, the residual radiation comes mostly in the form of fallout from the mushroom cloud.

5.1.1.5 Fallout

The final column presents the radius of 50% mortality from fallout 1 hour after the explosion. Of all of the threats described, fallout is the hardest to predict because of the influence of local, regional, or even global weather patterns. The mushroom cloud can rise into the atmosphere as far as 80,000 feet, where wind and rain influence the time and location for fallout to occur.[2] Individuals several miles from ground zero and well outside any radius presented in Table 5.1 can receive significant or even lethal radiation doses from fallout. However, while the air blast, thermal burns, and initial radiation are threats in all directions, fallout is a threat downwind from ground zero. Wind speed and direction vary at different altitudes, and it is safest to assume that fallout is a potential threat in all directions from ground zero. Individuals outside the blast zone generally will have several minutes to an hour or more to seek shelter before fallout arrives.

Radiation levels from fallout decrease with time in a relationship described by the 7/10 rule: for every seven-fold increase in the hours after a nuclear explosion, the radiation levels in fallout drop by a factor of 10.[3] The relative drop in fallout radiation levels using the 7/10 rule is shown in Table 5.3.

Using the 7/10 rule, 7 hours after an explosion, radiation levels will have dropped by a factor of 10. After $7 \times 7 = 49$ hours (about 2 days), the levels will have dropped by a factor of 100. After $7 \times 7 \times 7 = 343$ hours (about 14 days), the levels will have dropped by a factor of 1000. Most of the intense radiation will be gone within the first few days after the explosion. It is assumed that after 2 days, individuals can surface to quickly gather resources (Table 5.3 also shows how long an individual has to seek shelter, gather resources, etc.). Ideally, movement of emergency responders and sheltered individuals will be controlled until *measured* radiation levels are acceptable and evacuation instructions are provided. However, as a rule, it is best to stay sheltered for at least 14 days, after which time it is assumed that radiation levels will have dropped enough to be able to leave the area — quickly.

5.1.1.6 Combined Effects

A distance of 0.25 miles represents the radius of complete destruction for a 10-kiloton nuclear weapon (see Table 5.1). If an individual is just outside the 0.25-mile radius after a 10-kiloton explosion, the combined effects from all threats will likely be lethal. Chances of survival increase with distance from ground zero, but a victim does not have the luxury of choosing relative distance when an attack occurs. However, exposures to the intense heat, initial radiation, and fallout can be limited and thus increase chances for survival and decrease long-term health effects.

Figure 5.1 illustrates the primary threats based on the explosion of a 10-kiloton nuclear weapon. The distances are taken from Tables 5.1 and 5.2 and are intended to illustrate the impact such a weapon would have if detonated in a populated area. No specific fallout radius is provided because it depends on yield, location, and meteorological conditions and is very difficult to predict.

TABLE 5.3
Fallout Radiation Levels Using 7/10 Rule

Time after Fallout Deposition	Radiation Levels	Time to Receive 10 rem
0 hours (0 days)	1000 rem/hour	36 seconds
7 hours (0.3 days)	100 rem/hour	6 minutes
49 hours (2 days)	10 rem/hour	1 hour
343 hours (14.3 days)	1 rem/hour	10 hours

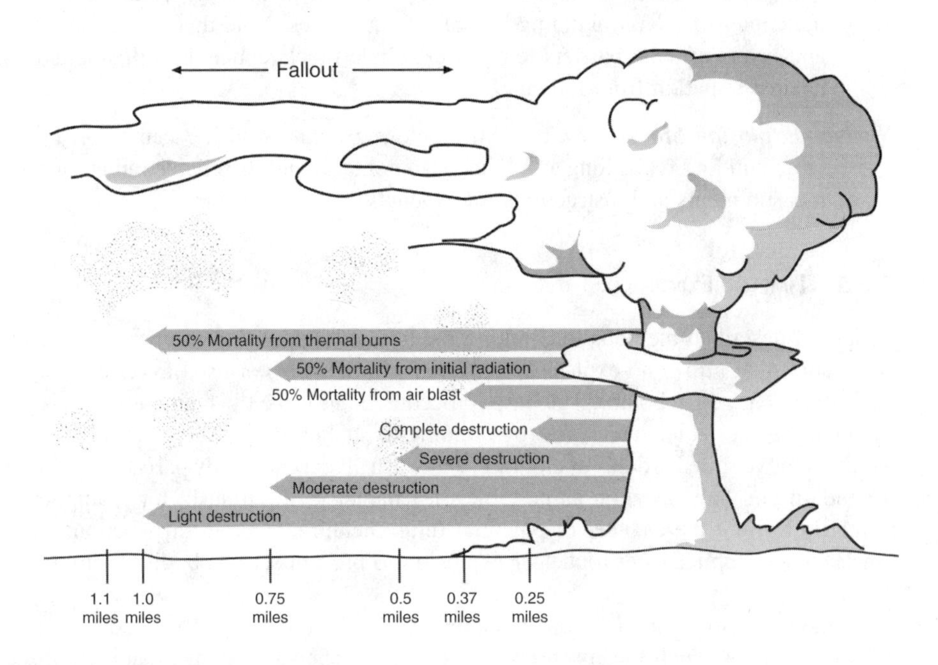

FIGURE 5.1 Primary threats from a 10-kiloton nuclear weapon.

5.2 RESPONSE TO NUCLEAR EXPLOSION

This section describes what an individual can do to limit exposure to intense heat from the fireball, initial radiation, and fallout after a nuclear explosion. This guidance is similar to that provided in Chapter 4. Time, distance, and shielding are the basic factors used to limit exposures and increase the chances of survival. The primary difference here is in the speed with which the three protective factors are implemented. Another difference is the order in which the factors are presented. Chapter 4 addresses time, distance, and then shielding. This chapter covers time, shielding, and then distance. The order of the factors was changed because individuals may not have the luxury of applying the distance factor until well after the explosion. As a result, the distance factor is discussed last. A few general nuclear explosion survival rules are as follows:

Nuclear Explosion Survival Rule 1: Close your eyes and turn away from the initial explosion; the intense heat from the fireball can cause temporary or permanent blindness.

Nuclear Explosion Survival Rule 2: Seek shelter or retreat behind a solid barrier immediately and stay hidden for at least 2 minutes. Note that every fraction of a second out in the open increases exposure to the intense heat, initial radiation, and air blast.

Nuclear Explosion Survival Rule 3: Get below ground if possible or move to the center of a large building before fallout arrives. Note that the ground and layers of solid barriers (e.g., brick, cinderblock, concrete) will shield against radiation from fallout.

Nuclear Explosion Survival Rule 4: Stay below ground or in the center of a large building for as long as 14 days. Do not leave unless reliable radiation measurements and instructions for evacuation are provided.

5.2.1 Time of Exposure

Chapter 4 described methods for limiting the time of exposure to weapons of mass destruction that utilize no explosives (e.g., aerosol delivery) or use of conventional explosives (e.g., dirty bomb). The basic procedure is to leave the contaminated area as quickly as possible, enter a nearby building to shelter against airborne contamination, remove soiled articles of clothing, and wash all exposed body parts (including the mouth and hair) as soon as possible. In Chapter 4, the time factor is applied primarily to limit the chances of potential future health effects. In this section, the time factor is applied after a nuclear explosion to prevent serious bodily harm and death.

Time is precious for anyone within a few miles of ground zero. Such individuals have no time to watch the mushroom cloud, gather personal items, calculate the distance from ground zero, or estimate the weapon's yield. As a general rule, if an individual can see a mushroom cloud, he is exposed to the initial radiation and heat.

Thus, every fraction of a second in the open increases radiation dose and the likelihood of serious thermal burns. The instant an individual realizes that a nuclear explosion has occurred, he should place as much solid material between his position and the rising fireball. The solid material can be a concrete or brick wall, deep ditch, building (preferably not constructed of glass), or anything that can act as a shield against radiation and heat.

The initial goal is to prevent serious thermal burns and high radiation doses. It can be accomplished by simply moving behind a wall for the first few minutes after the explosion. Ideally, a bomb shelter would provide the best protection. However, a basement, cellar, or a subway tunnel may be all that is available. If a suitable long-term shelter is not immediately available, wait several minutes in a temporary shelter until the initial heat, intense radiation, and air blast pass, then quickly find a more suitable long-term shelter (perhaps a neighbor's basement or nearby building) before fallout begins to cover the landscape. If possible, obtain whatever food, water, and emergency supplies are readily available before heading to the long-term shelter. Remain in the shelter until authorities confirm it is safe to evacuate. If you have no way to send or receive information (cell phone or radio), plan to stay in the shelter as long as 14 days.

An individual located well away from ground zero may be tempted to flee the area a few minutes after the explosion (refer to Tables 5.1 and 5.2 for relative distances) because it could take several minutes to an hour or more for the fallout to arrive downwind. However, the last place to be when fallout arrives is trapped in the open or in a traffic jam. Everyone else may try to escape too, and that will create traffic congestion. Staying in a designated or even a makeshift shelter is better than being stranded in an open area where chances of survival are severely diminished. Regardless of location relative to ground zero, the best procedure is to stay in a protected shelter until authorities confirm it is safe to evacuate.

Individuals should be prepared to stay in the underground shelter long enough for fallout radiation levels to subside. Two options are available for determining the appropriate time to leave the shelter: (1) obtain information from a radio or other source of outside communication or (2) prepare to stay underground for 14 days. The radio option is obvious, as emergency messages providing advice and evacuation information will likely be transmitted after a nuclear explosion. Assuming that no radio or other communication device is available, the 14-day rule may be used based on the relationship of fallout radiation levels and time after explosion (see Table 5.3).

Recall that for every seven-fold increase in the hours after a nuclear explosion, the fallout radiation levels drop by a factor of 10. Using the 7/10 rule and the information in Table 5.3, an individual can estimate how much time is available to find a more suitable shelter or gather more resources. Some radiation exposure may have occurred at the time of the explosion and inside a shelter. Therefore, individuals should take care to limit all potential exposure, for example, when gathering resources, to as little as reasonably achievable.

In summary, from the instant that a nuclear explosion occurs, an individual should take protective cover from the initial radiation and intense heat from the fireball. It will take a few minutes for the initial radiation and heat to subside. The next step is to locate in a suitable long-term shelter, preferably underground. Be prepared to stay in the shelter as long as 14 days or until authorities confirm it is safe to evacuate.

5.2.2 Shielding

Time and shielding can be merged into a single factor. The shelters described in Section 5.2.1 (walls, basements, etc.) really serve as shields from radiation, heat, fallout, and even from the air blast and flying debris. At the moment of explosion, radiation and heat travel at the speed of light and expose unshielded victims. At the instant of realization that a nuclear weapon has exploded, an individual should move as quickly as possible to a location behind a rugged shielding material.

Shielding material can be brick, concrete, steel, wood, or even the earth. Ideally, the shielding material will not collapse from the explosion or air blast or burn from the heat. If available, an underground shelter is the best option to avoid the air blast, thermal burns, initial radiation, and fallout. However, most victims do not have time to be terribly selective about where to seek shelter. The innermost rooms of a building may be the only shelter available. While they do not provide the same level of shielding as an underground basement, they provide more protection than an open environment. Seek the best shield in the immediate vicinity and stay there at least several minutes until the initial radiation and heat subside.

After the initial radiation and heat subside, the next challenge is to find a permanent shield to protect against fallout. Hiding behind a brick wall may provide protection for a few minutes, but will do little to protect against fallout. The best option is to move as deep underground as possible or to the middle of a building if no underground shelter is available. Keep in mind that below-level parking areas or middle floors of high-rise buildings and parking garages provide layers of concrete that shield radiation. Radiation from fallout can be significantly reduced by layers of building material or even by a foot of soil. Retreating underground or to the center of a building provides an easy (and hopefully quick) way to avoid radiation from fallout. After finding shelter, seek out the darkest corners and create a solid barrier between the hiding place and areas where radiation can penetrate (windows, doorways, corridors, etc.). The barriers can be stacked bricks, cinderblocks, sand bags, even water jugs, or wood. Even people can serve as shields. If there is a large group of people in a shelter, they should take turns spending time at the outside of the group. That way the dose is shared and not absorbed by a few individuals on the outer edges.

Figure 5.2 illustrates the effectiveness of an underground shelter to limit radiation dose. In the open, an individual will receive an unshielded radiation dose (D). If the individual is in a deep hole with only a thin overhead cover (e.g., plywood), the dose level drops by about a factor of ten (D/10). While this is better than being in an open area, fatal doses might still be received. If the cover is made of about one foot of concrete, the dose level drops by a factor of about 100 (D/100). Two feet or more of concrete will reduce dose levels by a factor of about 5000 (D/5000).[2]

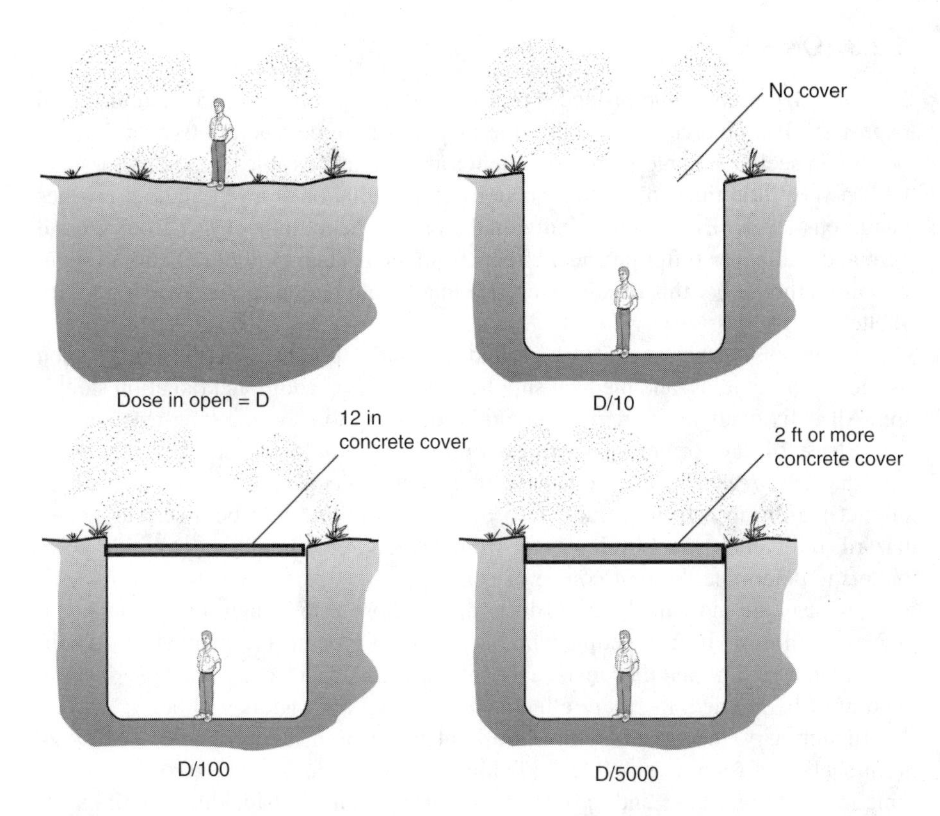

FIGURE 5.2 Using an underground shelter to limit radiation dose.

Although not as effective as a thick layer of concrete, multiple layers of other materials (e.g., a roof and several floors) may also be effective shields. For example, fallout will land on the roof of a multiple-story building. The roof and floors will keep the fallout away from individuals seeking shelter in inner rooms or a basement. Additionally, the multiple layers of wood, steel, and concrete help reduce dose levels in the shelter.

While underground, care should be taken to minimize the infiltration of contaminated air containing fallout particulates. Outdoor air should not be pumped into a shelter. A shelter should contain only recycled air or HEPA-filtered air if possible. If this is not an option, turn the heating and air conditioning system off. A positive pressure environment will force air through cracks and joints and keep contaminated air out. Positive pressure can be achieved by heating indoor air after making sure carbon monoxide poisoning is not a risk.

In summary, hide behind shields such as brick walls, cars, ditches, subway tunnels, etc. until the initial heat and radiation levels subside (after a few minutes). Then seek a more permanent shelter to shield against the radiation from fallout. Seek out dark corners and create additional shielding by stacking bricks, cinder-blocks, etc.

5.2.3 Distance

Section 5.1 presented the distances representing the levels of destruction and fatal exposures. If an individual is within the radius of light destruction (see Table 5.2), he should seek a suitable long-term shelter as soon as possible. He should assume he has very little time to get to a shelter before radiation from the fallout reaches dangerous levels. Even though individuals may be many miles away from ground zero and will never suffer physical effects from the heat and initial radiation, fallout remains a threat. For this reason, everyone in a fallout region should seek long-term shelter.

Individuals further away from ground zero have time to gather resources such as food, water, radios, and medical supplies and build up additional radiation shielding. All individuals in the region should be prepared to stay in a long-term shelter for at least 14 days or until authorities confirm it is safe to evacuate the area. Note that the term *region* is intentionally left undefined because it is very difficult to predict the distance from ground zero over which fallout will be a serious health hazard. Individuals many miles away from ground zero should use credible news outlets to determine the best course of action.

The average individual can do little in the short term to help the victims of a nuclear explosion. In fact, people who move toward ground zero in an effort to help can become victims and thus make a terrible situation even worse. The fewer people who must be rescued, the more effective emergency responders will be.

In summary, individuals within the light destruction radius should seek long-term shelter as soon as possible. Individuals further from ground zero have some time to gather resources and to build up additional radiation shielding. In all cases, individuals should plan to stay in their shelters as long as 14 days or until authorities confirm it is safe to evacuate.

5.3 SUMMARY

The five primary threats from a nuclear explosion are the destructive power of the nuclear explosion, the intense initial heat, the initial radiation, the air blast, and fallout. Almost nothing can be done to avoid the explosive power of a nuclear blast, but an individual can take some steps to avoid exposure to the remaining factors.

The intense initial heat lasts several seconds and the initial radiation lasts only a few minutes. Seek shelter behind some solid barrier (i.e., a brick wall or subway tunnel) as soon as possible to avoid high radiation doses and thermal burns. Hopefully, this solid barrier will also be able to endure the coming air blast. An air blast from a nuclear explosion travels at approximately 5 miles per second and carries with it glass, metal, or any debris in its path. By immediately retreating behind a solid barrier, individuals can avoid exposure to the initial heat and radiation and hopefully survive the air blast. The next challenge is seeking long-term shelter to avoid the radiation from fallout.

Fallout will occur at different times and in varying directions from the blast site, depending on weapon strength and meteorological conditions. It is best to assume that fallout will be a hazard anywhere in the region and at any time from a few minutes to an hour after the explosion. Individuals caught in the open when fallout arrives are at risk of receiving lethal exposure to radiation, so it is absolutely necessary to seek shelter as soon as possible. Individuals several miles from the destructive radius of the explosion have a short time to gather available food, water, and emergency supplies before retreating to a shelter (ideally, a shelter is already supplied). The best shelters are below ground with a couple of feet of concrete cover. However, individuals can still shield themselves from fallout by moving to the center of a large building or into a basement. Seal open spaces under doors, only use HEPA-filtered or recirculated air, and heat the air slightly if possible to create positive pressure. If this is not an option, turn the heating and air conditioning system off. Seek out dark corners of a shelter and build additional radiation shields with available resources (stack cinderblocks, sand bags, or any available material). Plan on staying in the shelter as long as 14 days or until authorities confirm it is safe to evacuate and an evacuation plan is provided. In summary, consider the following nuclear survival rules:

Nuclear Survival Rule 1: Close your eyes and turn away from the initial explosion because the intense heat from the fireball can cause temporary or permanent blindness.

Nuclear Survival Rule 2: Seek shelter or retreat behind a solid barrier immediately and stay hidden for at least 2 minutes. Note that every fraction of a second out in the open increases exposure to the intense heat, the initial radiation, and the air blast.

Nuclear Survival Rule 3: Get below ground if possible or move to the center of a large building before fallout arrives. Note that the ground and layers of solid barriers (brick, cinderblock, concrete) will shield against radiation from fallout.

Nuclear Survival Rule 4: Stay below ground or in the center of a large building as long as 14 days, and do not leave unless reliable radiation measurements and instructions for evacuation are provided.

REFERENCES

1. NCRP, *Management of Terrorist Events Involving Radioactive Materials*, National Council on Radiation Protection and Measurement Report 138, 2001.
2. Brodsky, A., Radioactive hazards in survival planning, *Radiation Protection Manage.*, 18, 34, 2001.
3. Shleien, B., *The Health Physics and Radiological Health Handbook*, Rev. Ed., Scinta, Silver Springs, MD, 1992.

6 Preparing for a Nuclear, Chemical, or Biological Attack

This chapter identifies measures that individuals living and working in a residential or urban setting should consider to protect against terrorist activities involving weapons of mass destruction. Specifically, these measures apply to residences, schools, churches, small businesses, large urban buildings, grocery and department stores and shopping malls, theaters and bowling alleys, hospitals, government buildings, airports, train and subway stations, and sports stadiums.

Facilities that have the highest likelihood of being targeted by terrorist activities are those that attract the largest density of people. For example, a school, large church, large urban building, grocery store, department store, shopping mall, entertainment center, hospital, airport, subway station, train station, or sports stadium (high public density areas) would be a much more likely target than a low public density area such as a residence or small business. For this reason, the emergency preparedness measures recommended for high public density areas are more stringent than those applicable to low public density areas. However, while low public density areas may not be likely targets for terrorist activities, they may be in close enough proximity to a high public density target to be impacted by contaminated smoke and airborne particulates. While post offices, courthouses, or police stations do not attract large groups of people, they are likely targets simply because they are government installations.

Table 6.1 presents recommended and optional emergency preparedness measures for a variety of residential and urban settings. The recommended measures should be seriously considered. Optional measures are not absolutely essential; they will provide added protection. In some settings, certain types of preparedness measures are not required. For example, a HEPA air filtration device is not required for a sports stadium.

The intent of Table 6.1 is to list practical measures that would be useful in the event of a terrorist attack. In many cases, the recommended emergency preparedness measures may already be in place to protect against natural disasters or other emergency situations.

It is understood that the ability to implement specific measures for a particular setting will be dependent on the availability of funding. For this reason, readers should use common sense based on knowledge of local environment and vulnerabilities when selecting specific measures to be taken. To ease the financial burden, the residential homeowner should consider purchasing items gradually. A business can temporarily divert overhead funding for nonessential activities, and a government installation could hold fundraisers or public meetings to discuss the need for additional funding to cover added protective measures.

The following sections provide specific details on the emergency preparedness measures identified in Table 6.1. They also provide names of vendors, Internet addresses, and telephone numbers. The guidance provided in the following sections should be used to evaluate (and supplement as necessary) existing emergency preparedness measures. It can also be used to begin to build emergency preparedness capabilities.

6.1 EMERGENCY PREPAREDNESS PLAN

All schools, large urban buildings, grocery and department stores, shopping malls, entertainment centers, hospitals, government buildings, airports, subway and train stations, and sports stadiums are strongly encouraged to develop or to amplify existing emergency preparedness plans to address the appropriate responses to terrorist attacks involving weapons of mass destruction. The purpose of the plans is to identify the specific measures that must be taken before, during, and after such incidents. Having an emergency preparedness plan in place prior to an emergency and providing training to staff on following the plan will ensure that everyone knows his or her specific roles and responsibilities.

Residences, churches, and small businesses may also want to consider developing emergency preparedness plans. While a plan for a residence would be less formal than a plan for a business or government building, having a plan in place will ensure that a family knows how to respond in the event of an emergency, including a terrorist attack. An emergency preparedness plan for a residence should be easily understood by even the youngest individual, and should address responding to different types of emergencies both at home and away from home. Occupants of smaller facilities should consider calling local emergency responders (fire department or police) for specific advice and literature on emergency preparedness planning.

An emergency preparedness plan should address the following key topics:

- Types of emergencies that could be encountered
- Specific roles and responsibilities during an emergency
- Methods to advise of an emergency
- Methods to differentiate various types of emergencies
- Types of alarm systems installed
- Methods of setting alarm systems
- Frequency for testing alarm systems
- Circumstances under which the building should be evacuated
- Circumstances under which occupants should remain inside the building
- Where to go, route to follow, and how long to stay
- Posting of maps showing evacuation routes
- Requirements for the building air purification system
- Requirements for the building water purification system

TABLE 6.1
Emergency Preparedness Measures for Residential and Urban Settings

	Residences	Schools	Churches	Small Businesses (e.g., bank)	Large Urban Buildings	Grocery and Department Stores and Shopping Malls	Entertainment Centers	Hospitals	Government Buildings	Airports, Subways, and Train Stations	Sports Stadiums
Emergency preparedness plan	○	●	○	○	●	●	●	●	●	●	●
Emergency preparedness training	○	●	○	○	●	●	●	●	●	●	●
Emergency preparedness practice drills	○	●	○	○	●	●	●	●	●	●	●
Alarm system	●	●	●	●	●	●	●	●	●	●	●
HEPA air filtration system	○	○	○	○	○	○	○	●	○	○	NR
Water purification system	●	●	●	●	●	●	●	●	●	●	●
Dust mask	○	○	○	○	○	○	○	○	○	○	○
Respirator	○	○	○	○	○	○	○	○	○	○	○
Protective clothing	NR	NR	NR	NR	NR	NR	NR	●	●	●	NR
Safety gloves	○	○	○	○	○	○	○	●	●	●	○
First aid kit	●	●	●	●	●	●	●	●	●	●	●
Cellular telephone	●	●	●	●	●	●	●	●	●	●	●
Radio	●	●	●	●	●	●	●	●	●	●	●
Emergency lighting	○	●	○	○	●	●	●	●	●	●	●
Emergency food supply	●	●	○	○	○	○	○	●	○	○	NR
Screening instruments	NR	○	NR	NR	○	○	○	●	●	●	○

● Strongly recommended
○ Optional
NR Not required

- Requirements for personal protective equipment
- Requirements for first aid kits
- Requirements for communication devices
- Requirements for emergency lighting
- Requirements for emergency food and water supplies
- Requirements for screening instrumentation
- Methods for identifying missing occupants
- What to do and what not to do
- Locations of other types of supplies and instructions on how they are to be used

Copies of the emergency preparedness plan should be distributed for all personnel to review. The emergency preparedness plan should be posted in building gathering places such as kitchens, break rooms, and lobbies. A map showing building evacuation routes should be posted throughout the building, particularly near doorways and exits.

6.2 EMERGENCY PREPAREDNESS TRAINING

Employees of schools, large urban buildings, grocery and department stores, shopping malls, entertainment centers, hospitals, government buildings, airports, subway and train stations, and sports stadiums should receive initial training and annual refresher training on the contents of the emergency preparedness plan. Contracting an emergency response specialist to assist with the preparation of the emergency preparedness plan and provide staff training is an option that should seriously be considered.

The emergency response specialist should preferably be an industrial hygienist, health physicist, medical physicist, or safety engineer with experience working in radiological, chemical, and biological environments. From the perspective of protecting staff from hazards associated with terrorist activities involving weapons of mass destruction, training should address the following topics at a minimum:

- Summary of the contents of the emergency preparedness plan
- General description of the various types of weapons of mass destruction
- Routes by which one may be exposed to contaminants released from these weapons
- Potential health impacts from chronic or acute exposure to these contaminants
- Ways to minimize exposure to these contaminants
- A discussion and demonstration of the alarm systems present in the building
- Appropriate response to each type of alarm
- When it is appropriate to evacuate the building
- When it is appropriate to seek shelter inside the building
- Where the evacuation maps are posted within the building

- What the evacuation routes are
- Details on the air circulation system of the building
- Options to reduce contamination pulled into the building (air circulation system in recycle mode)
- Person responsible for controlling the air circulation system after a terrorist attack
- Details on the water filtration system in the building and the types of contaminants it will remove
- Other ways to minimize contaminants allowed into the building
- Methods for controlling the spread of contaminants already in the building
- Methods for treating victims exposed to contamination and addressing common medical emergencies such as burns, cuts, and heart attacks
- Locations and contents of first aid kits
- Locations of emergency communication devices
- Details about emergency lighting and emergency food supplies
- Training on the use of basic hand-held radiation and/or chemical screening instrumentation
- Staying informed of emergency response activities and knowing when it is safe to exit the building

Staff should receive annual refresher training on the emergency preparedness plan. Newly hired staff should be required to complete this training before starting work. Contracting a psychologist to provide information to staff about the best methods for remaining calm during a terrorist attack, how to deal with post-attack depression, and related matters is another option that may be considered.

Residences, churches, and small businesses may also want to consider providing residents or employees with some level of emergency preparedness training. For example, parents should consider reviewing the emergency preparedness plan with their families at least once each year. Training may include going through emergency scenarios and ensuring that emergency supplies such as fresh batteries are available and in order. The yearly training will be beneficial for responding to terrorist attacks and in the more likely occurrence of a natural disaster.

6.3 EMERGENCY PREPAREDNESS PRACTICE DRILLS

Regarding schools, immediately after teachers are trained on the emergency preparedness plan, they should train students on how to correctly respond to the various types of alarm systems (e.g., fire, radiation, and chemical) in the building. Teachers should use cassette tapes, compact discs, or videotapes to allow the students to hear the differences in the sounds of the various types of alarms.

It is critical that students and teachers can clearly differentiate the types of alarms because the required responses may be very different. For example, a fire alarm requires evacuation of the building. A radiation alarm will in most cases require individuals to remain inside. Mistakenly exiting the building in a serious radiological emergency could subject people to much higher doses of radiation than if they remained inside the building. One or more emergency drills should be performed the first week of each new school year to ensure that students and teachers know how to distinguish all types of alarms and know the appropriate response to each.

For large urban buildings, grocery stores, department stores, shopping malls, entertainment centers, hospitals, government buildings, airports, subway and train stations, and sports stadiums, emergency drills should be performed the first week after the issuance of the emergency preparedness plan and annually thereafter. Grocery stores, department stores, shopping malls, entertainment centers, and sports stadiums should perform emergency drills before or after business hours when customers will not be present. Performing an emergency drill in an airport or subway or train station is more difficult because customers are often present 24 hours a day. As a result, practice drills in such facilities will likely need to be performed on a smaller scale (e.g., in a training center).

Residences, churches, and small businesses may also want to consider holding routine emergency response practice drills to ensure that family members and employees know how to implement the emergency preparedness plan.

6.4 ALARM SYSTEMS

All buildings (including residences) and sports stadiums require fire alarms or smoke detection systems. Schools, large urban buildings, hospitals, and government buildings should consider installing secondary alarm systems or intercom systems to alert occupants to hazards associated with weapons of mass destruction. When two alarm systems are installed in a building, one should be used to announce the evacuation of the building. The secondary alarm should be used to announce that personnel should remain inside the building until further notice. It is critical that the two alarm systems be easily distinguishable from one another, because responding incorrectly to either alarm could have life-threatening consequences.

Second alarm systems in churches, small businesses, grocery stores, department stores, shopping malls, entertainment centers, airports, and subway or train stations are *not* recommended, because the public will automatically assume that any alarm that sounds is a fire alarm and they will exit the building. For these types of facilities, an intercom system can be used to alert the public regarding the type of hazard present and how to respond. Using an intercom system instead of a second alarm system is also an acceptable option for schools, hospitals, and government buildings.

Some alarm systems are manually operated; others use sensors that automatically trigger the alarms. Smoke alarms installed in residences are examples of automatic alarms. A number of alarm systems on the market today will also sound automatically if a radiological hazard exceeds a specified threshold. Automatic radiation alarm systems provide a number of advantages over manually operated alarms:

- They only sound if radiation levels above ambient (background) levels are detected.
- They will likely provide personnel with early warnings of a radiation hazard.
- They avoid the possibility of accidentally pulling the wrong alarm.
- They eliminate prank alarms.

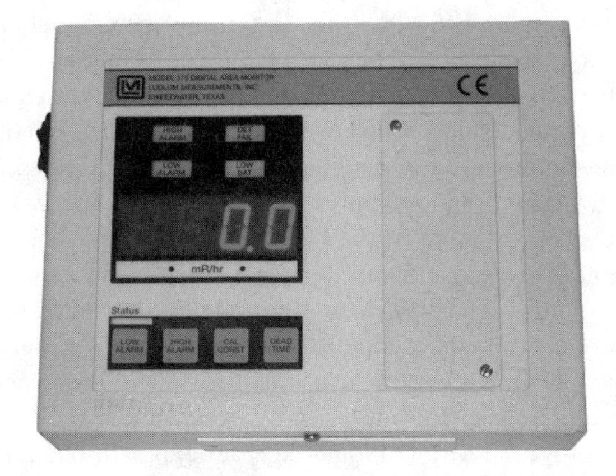

FIGURE 6.1 Automatic radiation-sensing alarm, Ludlum model 375/2.

The automatic radiation alarm system shown in Figure 6.1 (Part 48-2410, Ludlum Measurements, Inc., Sweetwater, TX, 1-800-622-0828, www.ludlums.com) is similar to those used at nuclear power plants. It contains radiation sensors that trigger the alarm to sound when activity levels exceed natural background levels. Installing several of these alarm systems around the perimeter of a school, large urban building, hospital, or government building would provide an early warning of a radiation hazard. For additional information, contact Ludlum Measurements (above) or Thermo Eberline (Santa Fe, NM, 1-505-428-3460) for details on other systems that may be considered.

Generally speaking, no practical and cost-effective automatic alarm systems are available to screen for chemical or biological warfare agents. Many chemical agents are very difficult to screen in the field because they are not volatile or require detection limits too low to be practical. Testing for chemical or germ warfare agents typically requires the collection of individual samples for laboratory analysis.

6.5 AIR PURIFICATION SYSTEMS

If a terrorist attack occurs near an occupied home or office building, the building can provide some level of protection from weapons of mass destruction, assuming its structural integrity is still intact and no imminent danger of fire is present. However, the purity of the air within the building must be maintained. This can be accomplished by:

- Switching the heating and air conditioning system to the recirculate mode to prevent pulling outside air into the building
- Turning the heating and air conditioning system off
- Using a HEPA air filtration unit (fitted with appropriate filters) to remove contaminants from the outside air before it is pulled into the building

Most buildings contain one of two major types of heating and air conditioning systems. The first type is designed for single family units that use 100% recirculated air. The second type is designed for commercial buildings with single or multiple air handling units with outside air intakes. For a system that does not include a fresh air intake, the fan should be set to run continuously during an emergency. This will recirculate the air through the system as long as the building has power and will maintain a positive air pressure in the building. If a system has outside fresh air intakes, they should be shut off and the fan set to continuous air recirculation. However, if the system does not include filtration of recirculated air, the best alternative is to shut off the system completely after sealing the outside air intakes. Chimney dampers, windows, pet doors, garage doors, and other openings should be shut as tightly as possible to prevent the entry of outside air.

The exception to these actions would be a building that has a HEPA air purification system specifically designed to remove contaminants from the fresh air intake. This type of filter system consists of pre-roughing filters, HEPA filters that can remove very small particles, adsorbent filters that can remove gases, and additional filters to further clean the air. Figure 6.2 shows a HEPA air filtration system. This system has the added advantage of being able to maintain positive air pressure in the building, effectively eliminating infiltration of outside air. Most systems also have provisions for emergency backup power. The only real disadvantage of a HEPA system is its high installation and maintenance cost. For this reason, HEPA systems are used primarily in hospitals and emergency control centers. Table 6.1 identifies air purification systems as "optional" for all building types except hospitals. Information supporting the design of HEPA air purification systems is available from American Air Filter International (AirShelter bag-in/bag-out multistate filtration system, Louisville, KY, http://www.aafintl.com/eprise/main/AAF_Intl/DustCollectors/bibo).

FIGURE 6.2 HEPA air purification system.

A number of different types of filters are available for single family units. Most are 1-inch diameter filters composed of materials such as fiberglass or pleated fabric. The pleated fabric filters have the highest efficiency of the passive filter types. One highly rated filter for standard air conditioning systems is the 3M Filtrete Ultra Allergen Reduction 1250 (Minnesota Mining and Manufacturing Company, St. Paul, MN, 1-888-364-3577, http://www.mmm.com/us/home_leisure/filtrete/411_ultra.jhtml). The cost per filter is $15 to $20 and a filter is effective for approximately 3 months. These and similar filters can be found at local hardware stores but are not effective for filtering smoke or very fine particulates. Filters that handle smoke and fine particulates are more expensive and generally require professional installation.

Residences or small businesses may consider installing one or more portable air purification systems (Figure 6.3), such as the CARE 2000 Air Defense System (CWR Environmental Products Inc., Glen Cove, NY, 1-800-444-3563, http://www.cwrenviro.com/care.htm). This system is often used in hospitals and health clinics to clean air of dust, pathogenic bacteria, viruses (large particulates), and small particulate chemicals, gases, molds, and pollens. The CARE 2000 air purifier would effectively remove radioactive dust particles from the air and should provide a sufficient germicidal dose to inactivate or kill smallpox and/or anthrax. Other portable air purification systems include:

- Austin Air Health-Mate (Austin Air Systems, Ltd., Buffalo, NY, 1-716-856-3700, http://www.austinair.com/)
- IQAir HealthPro and Cleanroom Series H13 (IQAir North America, Inc., Santa Fe Springs, CA, 1-800-500-4AIR, http://www.iqair.com/ENG/products/products.htm)

FIGURE 6.3 Portable air filtration unit.

TABLE 6.2
Comparison of Protective Measures for Airborne Contaminants

Protective Measure	Cost	Advantages/Disadvantages	
Standard air filters	$15 to $20	Advantages	Effective at improving daily indoor air quality; can be effective at filtering fine particulates; easy to install and maintain
		Disadvantages	Ineffective on very fine particulates and gases; must be replaced at least every 3 months; in an emergency, should only be used to filter recycled air; only provides protection indoors
Portable HEPA air filtration system	$300 to $1000	Advantages	Provides high level of protection for a single room with appropriate filters; affordable; provides protection from a variety of contaminants in the air (e.g., bacteria)
		Disadvantages	No benefit if electricity is interrupted; can purify air only in one room
Full-scale HEPA air filtration system	$10,000	Advantages	Provides high level of protection from airborne contaminants if used with appropriate filters; can be built into existing heating and air conditioning systems
		Disadvantages	May be too expensive for homeowners and small businesses; provides protection only indoors

To remain effective, an air purification system must be properly maintained and periodically tested. Table 6.2 summarizes the advantages and disadvantages of several options for removing airborne contaminants from the air.

Another option for insuring the quality of breathing air inside a building is installing a self-contained protective shelter such as the Bio-Shelter manufactured by Federal Group, Inc., 13281 Northend Avenue, Oak Park, MI, 248-545-5000, http://www.bio-shelter.com. The Bio-Shelter is a plastic-enclosed structure that can maintain positive or negative air pressure and is large enough to house four or five people along with required survival provisions. This shelter maintained in a positive-pressure mode will protect the inhabitants from surrounding contamination, and in a negative-pressure mode it can be used as an isolation unit.

6.6 WATER PURIFICATION SYSTEMS

Fresh water is an essential resource that must be available in the event of a terrorist attack. This section addresses options for maintaining fresh supplies of water that can be used for a few days or indefinitely. While a human can survive on 2 to 3 cups of water per day, assume for planning purposes that an individual needs 1 gallon of water per day. A family or business should have a minimum of several days' water supply for each family member or employee; however, a 2-week supply (14 gallons per person) is optimal. A larger water supply is needed for cooking and for pets.

The simplest and likely the least expensive way to maintain a fresh water supply is to purchase bottled water at a local grocery store. Stores often hold sales on bottled water or supplies can be purchased in bulk to reduce cost. Another option is purchasing a water filtration unit. Such units are available from several manufacturers and range from very simple filters that remove only particulates (e.g., radioactive particulates and anthrax spores) from the water, to units that remove everything including radioactive and other particulates (e.g., anthrax spores), metals, organics, herbicides, pesticides, bacteria, and parasites. The TGI 625U/DX (distributed by Pro Star Mechanical Technologies Ltd., Victoria, BC, Canada, 1-250-383-4558, http://www.prostar-mechanical.com/reverseosmosis.html) (Figure 6.4) and similar systems are examples of water filtration units that remove all these contaminants.

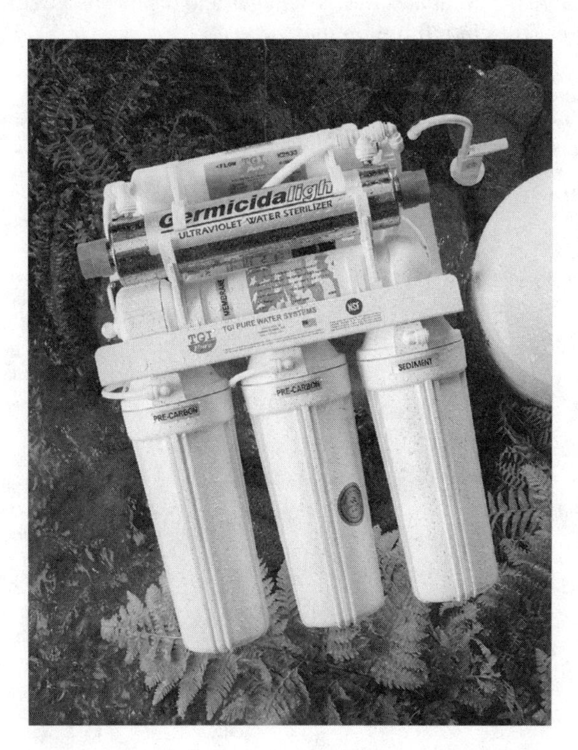

FIGURE 6.4 Water filtration unit.

Because a water filtration unit can be designed to fit most budgetary constraints, it is recommended that one of these units be installed in every home and building. Several companies that could be contacted for purchasing water filtration units are:

- Pro Star Water Purification Systems, 1-250-383-4558, www.prostar-mechanical.com
- Omni Water Filters, 1-877-420-9273, www.omni-water-filters.com
- Davnor Water Treatment Technologies Ltd, 1-403-219-3363, www.davnor.com
- Tiger Purification Systems Inc., 1-250-383-4558, www.watertiger.net

If a supply of water is not available when an emergency occurs, occupants of a home or building should quickly fill available reservoirs such as sinks, bathtubs, and other large containers. Time permitting, such containers should be cleaned before filling. A toilet tank — *not the bowl* — is another source of clean water. Natural sources such as streams or ponds can also be used, as long as the water is filtered before drinking. Water treatment tablets containing germicides are available for treating natural drinking water sources. Boiling water about 5 minutes is another effective way of eliminating organisms in stream or pond water, but water treatment tablets and boiling will not eliminate chemical or radiological contaminants. For this reason, it is recommended that water always be filtered to ensure that it is safe for drinking. Table 6.3 compares the advantages and disadvantages of storing bottled water versus investing in a filtration unit.

TABLE 6.3
Comparison of Fresh Water Sources

Protective Measure	Cost	Advantages/Disadvantages	
Bottled water	\$1 to \$3 per gallon	Advantages	Inexpensive; easy to store; can be purchased in bulk (e.g., 5-gallon containers); can be stored in cleaned 2-liter soda bottles
		Disadvantages	Shelf life is about one year
Water purification system	\$25 to \$1500	Advantages	Provides everyday supply of purified water; can be selected to remove as many (or as few) contaminants of concern
		Disadvantages	Filters must be replaced every 6 months, adding to initial expense; based on water flow through water lines; assumes flow will not be interrupted by a terrorist event

6.7 PERSONAL PROTECTIVE EQUIPMENT

Personal protective equipment is clothing and/or respiratory equipment worn to protect the body against various forms of contamination. Some of the most common forms of personal protective equipment include dust masks, air purifying respirators, protective suits made from particulates or chemically resistant materials such as Tyvek (E.I. Du Pont de Nemours & Company, Inc., Wilmington, DE) and/or other fabrics, and lightweight protective rubber gloves or chemical-resistant gloves.

If a dust mask is not available at the time of terrorist attack, breathing through a folded cloth (e.g., handkerchief or napkin) will provide some degree of protection against airborne particulates. A disposable dust mask affords a higher level of protection (Figure 6.5). The 3M Model 8210 N95 (3M, St. Paul, MN) claims to be 95% protective against airborne particulates ($\geq$ 0.3 microns in size), including radionuclide particulates. This type of mask is not fitted to the face (as a respirator is) and generally will provide an hour or so of heavy use before needing to be replaced. As the user's breath is trapped inside the mask, it quickly reduces its stiffness and potential seal to the face.

An improvement in disposable dust masks incorporates an exhalation valve and internal sealing surface. An example of this type of mask is the 3M Model 8511, which also claims 95% efficiency for particulates (Figure 6.6). Because this type of dust mask is effective in removing most airborne particles ($\geq$ 0.3 microns in size) from the breathing air and is relatively inexpensive, it is strongly recommended for use as an emergency preparedness measure. However, it should be noted that these masks will not provide protection against nonparticulate airborne contaminants (e.g., chemical warfare agents) and must retain their seal against the face in order to be effective.

It is recommended that each residence consider purchasing one or more dust masks for each family member. They are small enough to be inconspicuously carried in a purse, slipped into a coat pocket, stored in the glove compartment of a car, or stored in a house. Schools should consider purchasing dust masks for all staff

FIGURE 6.5 Disposable dust mask.

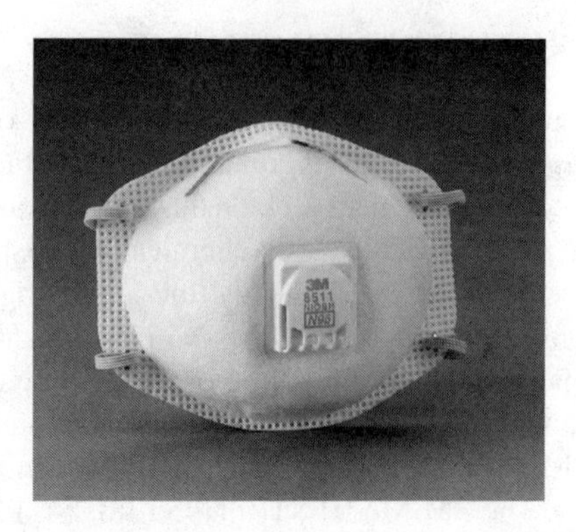

FIGURE 6.6 Dust mask with exhalation valve.

members and students. They could be stored in classrooms and distributed by teachers if needed. It is recommended that churches, hospitals, government agencies, and all of the other businesses listed in Table 6.1 should also consider purchasing dust masks for all their employees.

An emergency responder called to a terrorist attack should wear a half-face or full-face air-purifying respirator instead of a dust mask. While an air-purifying respirator is much more expensive than a dust mask, with the appropriate filters it can provide protection from a multitude of airborne contaminants including radioactive dust particles, volatile radionuclides, anthrax spores, and chemical warfare agents. The advantage of a full-face respirator over a half-face respirator is that the full-face respirator also provides eye protection and a better seal against the face. These types of respirators should be fitted to the wearer's face and the appropriate type cartridge selected for the agents of concern. However, this type of respirator produces resistance to breathing that makes it dangerous if the wearer has an upper respiratory or heart condition. Respirators and cartridges should always be stored in a cool place because the materials in the mask and cartridges can degenerate if exposed to harsh environmental conditions (e.g., in the trunk of a car).

Two examples of air purifying respirators are the MSA Ultra-Twin® full-face respirator (Mine Safety Appliances Company [MSA], Pittsburgh, PA), and the 3M Model 6800 full-face mask (3M, St. Paul, MN). Figure 6.7 shows an air-purifying respirator.

Other relatively inexpensive personal protective items that are not absolutely essential but should be considered are protective garments and gloves. One-piece coveralls with head covers and booties made from lightweight plastic such as Tyvek are relatively inexpensive, semi-repellent, and disposable. These types of suits are used in the nuclear and chemical industries to provide an added protection against contamination. Tyvek suits have sewn seams and are not recommended for chemical protection, except for vapors of low toxic solvents. Saranex-coated Tyvek or other heavy multilayer suits are required for long-term protection from chemicals. These heavier specialized suits significantly increase heat stress and should not be used by untrained personnel.

FIGURE 6.7 Air purifying respirator.

Latex surgeon's gloves are widely available in drug or department stores but are not good protective barriers against chemicals because they frequently have small holes and do not stand up to abrasion or tearing. Nitrile gloves are better for general protection against both particulates and chemicals. They are similar to latex gloves but provide much greater tear and puncture resistance. These unsupported single-layer gloves are relatively inexpensive, minimize loss of dexterity, and provide added protection to the hands when working in a contaminated environment. They can serve as undergloves when covered by heavy work gloves for work with potentially contaminated debris.

Dust masks, respirators, respirator cartridges, and other safety supplies can be purchased from local safety supply companies or through Internet web sites. Several large companies that market these items include:

- Aaron Industrial Safety Inc., 1-800-397-8871, www.aaronind.com
- MSA Safety Works, 1-888-672-4692, www.msasafetyworks.com
- R.J. Safety Company, Inc., 1-858-541-2880, www.rjsafety.com
- Brenton Safety, 1-800-733-4333, www.brentonsafety.com
- ABC Safety Mart, 1-800-646-5346, www.abcsafetymart.com

Safety professionals from these vendors can provide additional guidance on which dust masks, respirators, safety suits, and gloves are most appropriate for varying needs. They can also specify the exact respirator cartridges needed to protect against a specific list of contaminants of concern.

Because of varying body shapes and sizes, personal protective equipment must be fitted to each individual to ensure it will be protective. This is particularly important when purchasing a respirator, because a poor seal will significantly reduce protection. S

TABLE 6.4
Comparison of Personal Protective Equipment Options

Protective Measure	Cost	Advantages/Disadvantages	
Folded cloth	No cost	Advantages	No cost; nearly any cloth with a high thread count (e.g., folded handkerchief) can be used in an emergency; removes some portion of particulates from breathing air
		Disadvantages	For short-term use only; not designed to be a filter; one hand is always occupied; care must be taken to completely cover mouth and nose; not effective for removing fine particulates or gases
Dust mask	$1 to $10	Advantages	Inexpensive; easy to use; small enough to carry in small compartments; filters dust and particulates; some have two straps, an exhaust valve, and reinforced sealing surface
		Disadvantages	For short-term use only; not effective for removing very fine particulates or gases
Air-purifying respirator	$50 to $300	Advantages	Effective for removing up to 99% of very fine particulates and many gases; easy to use with proper training
		Disadvantages	Filters must be changed when saturated (see manufacturer's limits on filter use); no protection if used improperly; bulky and not practical to carry at all times; mask is usually made of rubber and must be kept in good condition (e.g., may crack if stored incorrectly); not effective in oxygen-deficient environments; must be fitted to be most effective; wearer must be medically qualified to wear a negative pressure respirator
Tyvek suit	$5 to $10	Advantages	Provides protection against alpha particles and low-energy beta particles; relatively inexpensive and disposable
		Disadvantages	Not impermeable to liquids; retains body heat
Latex or nitrile safety gloves	$0.50 to $3	Advantages	Provide protection against alpha particles and low-energy beta particles; resists some chemicals (depends on the type selected); relatively inexpensive
		Disadvantages	Latex gloves tear easily

Table 6.4 summarizes the advantages and disadvantages of several options for filtering airborne contaminants from the breathing zone and protecting the exterior of the body from contamination.

6.8 FIRST AID KITS

A well-stocked first aid kit is strongly recommended for all facilities listed in Table 6.1 and should include:

- Bandages (standard and butterfly types)
- Gauze pads (in multiple sizes)
- Gauze wraps
- First aid tape
- Sanitizing wipes
- Scissors
- Elastic wraps
- Arm sling
- Finger splints
- Eye pads
- Instant ice packs
- Antibacterial ointment
- Burn gel
- Eye flush
- Latex examination gloves
- Ibuprofen
- Vitamin supplements
- Other daily medications (e.g., pain relievers)
- Potassium iodine tablets (optional)

First aid kits are readily available at pharmacies and over the Internet. Homeowners may also want to consider maintaining supplies of practical home remedies and supplies such as:

- Cranberry juice (for bladder problems)
- Apple cider vinegar (for internal ailments)
- Tea tree oil (for exterior infections)
- Colloidal silver (a natural antibiotic)

Businesses should purchase first aid kits for the number of employees and type of business. Cost is about $1 per person, depending on the manufacturer, type, and number of kits purchased.

Potassium iodine tablets can be used to reduce radioactive iodine exposure to the thyroid gland. According to the National Council of Radiation Protection and Measurement (NCRP), taking 130 milligrams of potassium iodine at or before exposure to radioactive iodine effectively blocks nearly 100% of radioactive iodine from reaching the thyroid.[1] Waiting 4 hours after exposure to take potassium iodine will only block approximately 50% of radioactive iodine from reaching the thyroid; thus, if potassium iodine is to be taken, it should to be taken as soon as possible following radiation exposure.

Several companies to consider when assembling or purchasing first aid kits are:

- Leonard Safety Equipment, 1-800-556-7170, www.leonardsafety.com
- First Aid Supplies Online, 1-800-874-8767, www.firstaidsupplieson-line.com
- Major Safety, 1-800-446-8274, www.majorsafety.com

6.9 COMMUNICATION DEVICES

Because electrical lines may be knocked out by emergency events, it is recommended that one or more cellular telephones be maintained in each of the facilities identified in Table 6.1 to be able to call for help and communicate with family members. Communicating with rescue workers and family during an emergency event will help minimize some of the psychological effects that may be experienced. Each building should also maintain one or more battery-powered AM/FM or short-wave radios and several sets of replacement batteries. The radios can be used to obtain updated reports on emergency response efforts and other useful information.

6.10 EMERGENCY LIGHTING

It is recommended that multiple battery-powered light sources (and replacement batteries and bulbs) or other emergency lighting devices be maintained in all schools, large urban buildings, grocery and department stores, shopping malls, entertainment centers, hospitals, government buildings, airports, subway and train stations, and sports stadiums in the event that electrical power is unavailable for an extended time. Backup lighting helps emergency response workers see where they are going, allows hospital staff to continue tending to critically ill patients, and prevents panic. Emergency lighting is optional for residences, churches, and small businesses; one or more battery-powered camping-type lanterns or flashlights are usually sufficient for residences and small businesses.

6.11 EMERGENCY FOOD SUPPLIES

Residences, schools, and hospitals should maintain emergency food supplies that can feed family members, students, and patients for a minimum of several days and preferably for 2 weeks. Food should include nonperishable canned items such as tuna, soups, stews, vegetables, fruit, and fruit juices. Residence, school, and hospital kitchens should be supplied with disposable plates, bowls, cups, eating utensils (primarily spoons), multiple hand-operated can openers, and a large supply of plastic garbage bags and paper towels. Stocking an emergency food supply is optional for the other facilities identified in Table 6.1.

An average individual can survive for an extended period on about 500 calories per day, as long as he or she has plenty of water and physical activity is kept to a minimum. Five hundred calories are equivalent to approximately two 6-oz cans of tuna in water or one 15-oz can of a fruit or vegetable. For a family of three to survive 2 weeks, they must store at least 21,000 calories of nonperishable food or the approximate equivalent of 84 6-oz cans of tuna fish or 42 15-oz cans of fruits or vegetables.

Food tablets occupy little space and provide high-calorie dietary supplements. They can be used to "stretch" food stores over longer periods. Foods in dried or powdered form are also available and can be stored for long durations without spoiling. If dried foods are used for emergency supplies, it is important to ensure that containers are properly sealed to prevent insect infestation.

6.12 SCREENING INSTRUMENTS

It is strongly recommended that hospitals consider purchasing one or more hand-held radiation survey instruments (Figure 6.8) for the purpose of identifying radiological contamination. In the event of a terrorist incident, one or more radiation survey instruments in a hospital emergency room would allow doctors to determine whether arriving patients are radiologically contaminated. These instruments could provide doctors with information regarding the levels of radiation received and could help control the further spread of the contamination.

One particular radiation survey instrument that should be considered is the Ludlum Model 3 survey meter, in combination with the Ludlum Model 44-9 pancake Geiger-Mueller detector. This combination of meter and detector can scan for alpha-, beta-, and gamma-emitting radionuclides simultaneously and provide count-per-minute readings. Another instrument that may be considered is a dose rate meter. For information on additional radiation survey instruments refer to the Ludlum web page at www.ludlums.com. For details on more sophisticated radiation survey methods (e.g., robotic methods, pipe surveying, airborne survey methods), refer to Byrnes, M.E., *Sampling and Surveying Radiological Environments*, Lewis Publishers, Boca Raton, FL, 2001 (ISBN 1-56670-364-6). Training on the proper use of the selected radiation survey instruments should be provided during emergency preparedness training (Section 6.2).

A wide variety of commercial equipment is available for detection of hazardous chemicals, including a number of chemical warfare agents. For example, ion mobility spectroscopy is used to detect nerve, blister, and blood agents. The Chemical Agent Monitor is a portable, hand-held point detection instrument that uses ion mobility spectrometry to monitor nerve or blister agent vapors. However, minimum detection limits are approximately 100 times the acceptable exposure limit for nerve agents, and approximately 50 times the acceptable exposure limit for blister agents.

Acoustic wave sensors are also used to detect nerve and blister agents. The surface acoustic wave chemical agent detector (SAW Mini-CAD) is a commercially available, pocket-sized instrument that can monitor for trace levels of toxic vapors of sulfur-based mustard agents (e.g., distilled mustard) and G nerve agents (e.g., tabun, sarin, soman) with a high degree of specificity. Colorimetric tubes are the

FIGURE 6.8 Ludlum Model 3 survey meter with Model 44-9 pancake Geiger-Mueller detector.

most common detection technologies used by hazardous materials (HAZMAT) teams. A HAZMAT team's analytical capabilities usually include tests for chlorine, cyanide, phosgene gas, and organophosphate pesticides. Routine HAZMAT tests rarely include the capability to test for blister agents.

The M18 and M256A1 detection kits are military items. The M18 is a colorimetric device for measuring the concentrations of selected airborne chemicals. The M18 comes with detector tubes for cyanide, phosgene, lewisite, distilled mustard, tabun, sarin, soman, and V gas. The M256A1 kit can detect low concentrations of cyanide, blister, and nerve agents in vapor form. The tests take approximately 15 minutes to perform. The sensitivity of this kit is such that the tests may provide a negative reading at concentrations below those immediately dangerous to life and health (IDLH). The IDLH is the maximum concentration of a contaminant to which a person could be exposed for 30 minutes without experiencing any escape-impairing or irreversible health affects. M8 and M9 detection papers provide a rapid, inexpensive method to test for the presence of liquid mustard or nerve agents. They should be used only as a screening test due to the tendency of the papers to show false positive readings.

Additional detectors currently available use other technologies such as electrochemical detectors for blister, nerve, blood, and choking agents, and infrared spectroscopy detectors or photo ionization detectors for the detection of blister and nerve agents. Mass spectrometry, gas chromatography, and Fourier transform infrared spectrometry technologies are the bases for fixed laboratory agent detection. They provide definitive identification of agents and possess the sensitivity to detect low levels of the agents in the range of occupational exposure limits.

Government buildings, airports, subway stations, and train stations are encouraged to consider purchasing one or more radiation and chemical agent survey instruments, primarily for security purposes. The instruments could be used to screen suspicious packages or personnel to determine the presence of radioactive or chemical contamination. Survey instruments are optional for schools, large urban buildings, grocery and department stores, shopping malls, and sports stadiums, and are not required for the other structures identified in Table 6.1.

6.13 SUMMARY

Many emergency preparedness measures that should be taken to prepare for a terrorist attack may already be in place as a result of preparing for natural disasters or other emergencies. The guidance in this section should be used to evaluate (and supplement) existing emergency preparedness measures or can be used to begin building emergency preparedness capabilities. The following is a summary of measures that, along with guidance from local emergency responder groups (e.g., fire departments), can significantly increase the chances of surviving a large-scale terrorist attack.

An emergency preparedness plan defines roles and responsibilities, where to go, what to do, and who is in charge when an emergency situation arises. Employees and family members should be instructed in the proper execution of plans and should be routinely drilled (at least yearly) to identify vulnerabilities. Ask local emergency responder groups for assistance in developing plans and get their advice on conducting training and practice drills. Prepare for a range of emergency events including a terrorist attack.

Planning Summary Develop an emergency preparedness plan for emergency events including a terrorist attack. Make sure training and practice drills are completed at least yearly, and look for potential vulnerabilities so the plan can be optimized.

Warning devices may be used to identify potential hazards from a terrorist attack or to warn of an emergency situation. Radiation alarms can be used to identify contamination after explosion of a dirty bomb. Alarm systems can use different signals to identify specific emergency response procedures. A building intercom system and emergency lighting can also signal an emergency event and can help guide potential victims to safety. These warning devices can limit or eliminate exposures to hazardous substances.

Warning Device Summary Use warning devices such as radiation alarms, fire alarms, intercom systems, and emergency lighting to warn of potential hazards and identify appropriate emergency response procedures.

Emergency preparedness includes the installation of protective equipment and the collection and storage of resources. Protective equipment may include a HEPA air filtration and water purification system, and resources may include dust masks, respirators, protective clothing, safety gloves, first aid kit, cellular telephone, emergency lighting, food supply, and screening instruments. Store first aid kits in kitchens, lobbies, vehicles, and places where people gather or would seek shelter. Enough food and water should be on hand to last 2 weeks, or about 1 gallon of water per person per day, and enough nonperishable food to supply at least 500 calories per person per day. Air purification devices and protective clothing (dust masks, safety gloves, etc.) can help limit exposure to hazardous substances by filtering or otherwise limiting direct contact with contaminants. Finally, communication devices such as radios and cell phones may be used to receive vital information about potential hazards and evacuation plans.

Resources Summary Gather and store emergency resources including air and water filtration systems, dust masks, respirators, protective clothing, safety gloves, first aid kits, cellular telephones, emergency lighting, food supplies, and screening instruments.

REFERENCE

1. ICRP, *Recommendations of the International Commission on Radiological Protection*, ICRP Publication 26, International Commission on Radiological Protection, Stockholm, 1977.

7 Guidance for Emergency Responders

Emergency responses to incidents involving weapons of mass destruction are much different from responses to incidents involving conventional explosives. For example, while the debris from conventional explosives may be extensive and present hazards to the individuals responding, such as fires or structural instability, the materials are not inherently hazardous. In this situation, the site can be secured and forensic investigations can be conducted similarly to investigations at other crime scenes.

When an explosive device disperses radioactive, chemical, or biological materials, the treatment of casualties is more difficult because of the presence of contamination. Emergency responders may face a life-threatening situation unless appropriate precautions are taken. The affected area for an attack involving weapons of mass destruction will likely be much larger than the immediate scene of a conventional crime. The incident will be difficult to manage until appropriate monitoring equipment and well-trained technical responders arrive at the site. Forensic investigation will be complicated by the need to wear protective equipment, and evidence will likely be contaminated.

7.1 PRIORITIZING INJURIES

Treatment of life-threatening injuries should almost always take precedence over measures to address radioactive, chemical, or biological contamination or exposure. Injured individuals should be stabilized if possible and immediately transported to a medical facility. It is recommended that an individual with training in the areas of radioactive, chemical, and biological hazards accompany the first patients to the hospital and serve as an advisor to the medical team.

The possibility of contamination of patients may be determined in the field, en route to a treatment facility, or at a treatment facility, depending on the condition of the patients. The facility receiving the patients should be informed of the estimated number of casualties, the natures of their injuries, and details on any suspected contamination that may be present. Injured personnel should be sorted and treated according to standard medical guidelines. If possible, individuals suspected of being contaminated should be separated from other patients and receive preliminary decontamination prior to treatment (see Section 7.3 for decontamination procedures).

Individuals who have only received external contamination and are not otherwise injured should preferably be decontaminated at a location other than a hospital.

Patients who show no evidence of external contamination but may have received internal contamination via a wound or inhalation or ingestion of contaminants do not need to be decontaminated prior to treatment. However, blood, vomit, urine, and feces from these patients may be contaminated and should be controlled as such.

A patient with a large amount of radioactive material imbedded in a wound warrants special attention because the material could cause a significant exposure hazard to treatment personnel. Dose equivalent rates from fragments resulting from the explosion of a nuclear reactor may be as high as 100 rem per hour.[1] The symptoms that may be displayed by individuals exposed to weapons of mass destruction are presented in Tables 3.2 and 3.3.

7.2 ASSESSING PATIENTS FOR CONTAMINATION

Contamination assessment of an injured individual should be performed by a trained health professional under the supervision of on-scene medical personnel. This assessment should involve the collection of gross alpha, gross beta, and gross gamma radiation measurements using hand-held radiation survey instrumentation, nasal wipes, and collection of saliva, blood, and other samples for laboratory testing. Section 6.12 recommends one of many survey instruments that should be considered for performing radiological survey assessments.

If the patient is in a contaminated area, he should be moved to a clean area. Radiation surveys should be performed by passing a probe slowly over the entire surface of the patient's body. Care should be taken not to contaminate the probe by contact with the patient or any other potentially contaminated surface.

The distribution of contamination on the body and other relevant information (e.g., locations of wounds) should be recorded. Administrative information that should also be recorded includes the patient's name; the name of the individual conducting the survey; the time, date, and location of the survey, and the serial number and type of instrument. A survey form with a diagram of an anatomical figure (Standard Form 531) is available from the U.S. General Services Administration's web site (http://www.gsa.gov/forms/medical.htm) and is suitable for this purpose.

Examples of information that can be collected by medical and radiological health personnel at the scene or during transport to the hospital that may be helpful in the early medical management of contaminated cases are:

Circumstances of the incident:

- When did the terrorist event occur and what were the circumstances?
- What are the most likely pathways for exposure?
- How much radioactive, chemical, or biological material is potentially involved (use nasal smears to estimate radionuclide intake)?
- What injuries occurred?
- What potential medical problems may be present besides radioactive, chemical, or biological contamination?
- What measurements were made at the site of the incident (by air monitors, smears, fixed radiation monitors, nasal smear counts, and skin contamination levels)?

Status of patient:

- What radionuclides or chemical/biological agents now contaminate the patient?
- Where and what are the radiation measurements at the surface?
- Was the patient also exposed to external gamma radiation?
- If exposure information is available, what was learned from processing this data?
- What information is available about chemical and physical properties of the compounds containing the radionuclides (e.g., solubility and particle size)?
- What decontamination efforts, if any, have been attempted?
- Were they successful?
- What therapeutic measures (e.g., the use of blocking agents or isotopic dilution procedures) have been taken?

Patient follow-up:

- Has any clothing removed at the site of the incident been saved?
- What excreta have been collected?
- Who has the samples?
- What analyses are planned?
- When will analyses be completed?

7.3 PERSONNEL DECONTAMINATION PROCEDURES

External decontamination procedures are designed to minimize internal contamination of patients and the individuals providing care. Radionuclides on the intact skin surface rarely produce a high enough absorbed dose to be hazardous to patients or to medical staff. However, this is not the case for chemical hazards. Prior to implementing decontamination procedures, a patient should be moved to an upwind position outside of the area of contamination.

Under circumstances in which a large number of individuals must be decontaminated, they should be transported to suitable locations (e.g., sport centers or military installations) where large shower facilities are available or to a temporary outdoor facility (in good weather conditions) organized to accommodate this type of procedure. In some cases, the authorities might consider issuing guidance for people to shower in their own homes while waiting to be evacuated.

7.3.1 Radiation Decontamination Procedure

During radiological decontamination efforts, caregivers should consider the chemical nature of the contaminant, the medical needs of the patient, the seriousness and extent of contamination, and the presence of wounds. Prior to implementing decontamination procedures, remove all outer clothing from the contaminated individual and place the clothing in a sealed container such as a plastic bag. Save clothing so that samples can be collected at a later time. The order of decontamination should proceed as follows: 1) wounds, 2) body orifices, and 3) intact skin.

The following steps are usually effective for decontaminating wounds:

1. Place drapes around the wound.
2. Cleanse the wound using standard wound cleansing procedures and mild-cleansing agents (e.g., rinsing with saline or distilled water).
3. Blot dry.
4. Remove drapes and survey.
5. Repeat as necessary (noting that scabs may contain contamination).
6. Dress wounds with a waterproof cover before proceeding to cleanse other areas.

These steps may produce "runoff" that has to be managed. Channeling runoff into a receptacle or using absorbent materials can minimize the spread of contamination.

Body orifices should be decontaminated next using swabs and irrigation. A Morgan lens or nasal cannula (small tube used for draining body cavities) may be used to decontaminate the eyes. Again, runoff should be controlled to prevent the spread of contamination.

Skin decontamination is performed last and generally begins with the least aggressive techniques and the mildest cleansing agents. The following steps are usually effective for intact skin decontamination efforts:

1. Drape clean areas.
2. Wipe towards the center of contaminated areas with clean absorbent materials (e.g., cloth) — motion is like cleaning up a paint spill.
3. Blot to dry.
4. Repeat as necessary.

Additional steps may include flushing with soap and water or showering starting with head and working down. Care should be taken to keep contaminated materials away from the eyes, nose, and mouth. Care should also be taken to avoid causing mechanical, chemical, or thermal damage to the skin. It should be noted that complete decontamination is generally not possible because some radiological contaminants can remain fixed to the skin surface. The objective of the decontamination effort is to remove the majority of the external contamination.

7.3.2 Chemical Agent Decontamination Method I

The following 10-step process is recommended for decontaminating personnel exposed to distilled mustard (HD), arsenicals (MD, PD, ED, and L), nitrogen mustards (HN-1, HN-2, and HN-3), oximes (CX), and nerve agents (GA, GB, GD, and VX):

1. Remove all signs of obvious gross contamination from patient by scraping with a wooden stick or by sweeping or blotting matter away. Absorbent material such as earth, flour, or dry detergent soap powder can be applied to limit the amount of chemical traveling through the skin.
2. Remove all outer clothing and garments rapidly, being careful not to contaminate skin. If clothing cannot be removed easily, it should be cut away using scissors. Place clothing in a sealed container such as a plastic bag.
3. Remove all other items on the body such as rings, necklaces, watches, hearing aids, glasses, contact lenses, and artificial limbs.
4. Wash hands with copious amounts of soap and water, rinsing thoroughly.
5. Flush patient's eyes with copious amounts of water.
6. Wash patient's face and head (including hair, beard) with soap and water and rinse thoroughly.
7. Apply 0.5% bleach solution gently by blotting or using hands to apply over all areas of the skin starting from the neck down. Do not rub solution too vigorously as this may irritate skin. If bleach solution is not available, go directly to Step 8.
8. Thoroughly wash the entire body with soap and water, rinse with copious amounts of water.
9. Change patient into fresh clothing or wrap in blankets/towels.
10. Soak items removed from the body (Step 3) in full 5% hypochlorite solution for 5 minutes, then rinse thoroughly prior to returning them to patient.

7.3.3 Chemical Agent Decontamination Method II

If personnel have been exposed to blood agents (AC, CK, and SA) or choking agents (chlorine, CG, or DP), the following decontamination method is recommended prior to treatment:

1. Remove all outer clothing and garments rapidly, being careful not to contaminate skin. If clothing cannot be removed easily, it should be cut away using scissors. Place clothing in a sealed container such as a plastic bag.
2. Remove all other items on the body such as rings, necklaces, watches, hearing aids, glasses, contact lenses, and artificial limbs.
3. Wash hands with copious amounts of soap and water, rinsing thoroughly.
4. Flush patient's eyes with copious amounts of water.
5. Wash patient's face and head (including hair, beard) with soap and water and rinse thoroughly.
6. Thoroughly wash the entire body with soap and water (or dilute hypochlorite solution), rinse with copious amounts of water.
7. Change patient into fresh clothing or wrap in blankets/towels.

7.4 EXPOSURE GUIDANCE FOR EMERGENCY RESPONDERS

Special guidance for exposures often in excess of dose limits is required for emergency response operations. In severe disasters, prompt but well considered actions can potentially save lives and avert significant harm to the public. NCRP Report 116 titled *Limitations of Exposure to Ionizing Radiation*[2] provides broad guidance for emergency responders.

Normally, only actions involving saving human life justify acute exposures significantly in excess of the annual effective dose limit. The use of volunteers for exposures during emergency actions is desirable. Older workers with low lifetime accumulated effective doses should be chosen from among the volunteers whenever possible. Exposures during emergency operations that do not involve lifesaving should, to the extent possible, be controlled to the occupational dose limits. Where this cannot be accomplished, it is recommended that (from a radiological perspective) a limit of 50 rem effective dose and an equivalent dose of 500 rem to the skin be applied, which is consistent with International Commission on Radiological Protection (ICRP) recommendations.[3]

When, for lifesaving or equivalent purposes, the equivalent dose approaches or exceeds 50 rem, workers must understand the potential for acute effects and the increase in lifetime risk of cancer. If internally deposited radionuclide exposures are also possible, these types of exposures should be taken into account.

Workers should also be selected based on their experience in performing required emergency tasks, as the time to accomplish a task will likely be reduced and help minimize worker exposure. The number of workers included in such tasks should be kept as low as strictly necessary for the tasks to be carried out. Only nonpregnant workers over the age of 18 should be selected.

Because of the great uncertainty in estimating exposures during the early phase of the incident and the recognized need to control exposure during potentially sensitive gestational periods, it is prudent for minors or responders who are pregnant or potentially pregnant to volunteer for service during a later, more controlled phase of the incident.

Unless an incident occurs at a nuclear facility, it is possible that the first responders will not recognize the radiological aspects of the event. Because it is not likely that all responding individuals will have received training normally required of workers who are routinely occupationally exposed, it is necessary to establish a mechanism to ensure that the workers are not likely to receive an unacceptable level of exposures, while at the same time allowing them to perform critical missions during the early phase of a disaster.

For this reason, NCRP recommends that emergency response personnel and first response vehicles should be equipped with radiation detection equipment to alert them for a radiologically compromised environment. Furthermore, this equipment should be designed to alert the responders when unacceptable ambient dose rates or dose limits are reached. Responders should wear appropriate personal protective equipment (e.g., chemical safety suits, respirators).

Emergency response personnel assigned to respond to a scene with this equipment should receive training that includes the operational characteristics of the equipment, the operational quantities to be measured, and the risks associated with exposures that correspond to the preset levels of the alarms.

NCRP recommends that an ambient dose rate of approximately 10 millirem per hour is a suitable alarm level. This value is significantly higher than natural background levels so that false positive indications are avoided, but not so high that an emergency responder is likely to receive exposure that would approach the annual limit for a member of the general public if exposed in areas below this value. It is also an ambient dose rate at which it is appropriate to establish an initial control point to restrict access for radiological control purposes to individuals who are not necessary at the scene. Initial responders should not proceed beyond the point at which the alarm level has been reached unless they have compelling reasons to do so, such as rescuing injured persons or performing time-sensitive actions to regain control of the scene. However, if the first responders include personnel with radiation health expertise and more sophisticated equipment, it is more appropriate that they make judgments involving higher exposures at the scene, taking into account all relevant factors specific to conditions at the scene.

7.5 TRAINING FOR EMERGENCY RESPONDERS

Training of emergency responders (firefighters, police, and emergency medical services) for responding to weapons of mass destruction should emphasize critical concepts for self-preservation and effective casualty management. Training should include both classroom instructions on basic concepts and principles, hands-on demonstrations of required skills, and drills to reinforce basic procedures. The recommended topics to be covered include:

- Defining potential incidents involving radioactive materials and chemical and biological agents
- Identifying chemical properties of radioactive materials and chemical and biological agents
- Defining general radiation protection principles:
- As low as reasonably achievable (ALARA)
- External exposure (time, distance, and shielding)
- Internal exposure (respiratory protection, hygiene, and monitoring)
- Defining how humans are exposed to radioactive materials and chemical/biological agents
- Identifying how exposure to radioactive materials and chemical and biological agents can impact human health, human genetics, or an unborn fetus
- Identifying various ways to detect the presence of radioactive materials and chemical/biological agents
- Defining relevant emergency procedures (including reference or action levels)

- Demonstrating use of radiation survey instruments and instruments for screening chemical agents
- Demonstrating sample collections methods for chemical and biological analysis
- Discussing contamination control and decontamination procedures
- Managing casualties
- Discussing command, control, communication, and coordination

Refresher training should be provided annually at a minimum. New hires should be required to complete the training before beginning work.

REFERENCES

1. NCRP, *Management of Terrorist Events Involving Radioactive Materials*, Report 138, National Council on Radiation Protection and Measurements, Bethesda, MD, 2001.
2. NCRP, *Limitation of Exposure to Ionizing Radiation*, Report 116, National Council on Radiation Protection and Measurements, Bethesda, MD, 1993.
3. ICRP, *1990 Recommendations of the International Commission of Radiological Protection*, Publication 60, Pergamon Press, New York, 1991.

8 Summary of Recommendations

This chapter summarizes the primary recommendations provided in this book. Following these recommendations will help emergency responders minimize their exposures to various types of weapons of mass destruction.

8.1 MINIMIZING EXPOSURE TO RADIATION (DIRTY BOMB) AND WARFARE AGENTS

The following rules provide guidance on how emergency responders can best minimize exposure to weapons of mass destruction that utilize no explosives (e.g., aerosol delivery of agent) or use conventional explosives (e.g., dirty bomb):

Time Rule 1: For individuals in the vicinity of a terrorist attack, stay inside or move inside an undamaged building as quickly as possible and stay there until authorities confirm it is safe to evacuate.

Time Rule 2: For individuals at the site of a terrorist attack, leave a damaged building or affected area in a quick and orderly manner and seek shelter in a nearby, preferably undamaged, building following an emergency response plan if one exists.

Time Rule 3: Minimize the time of exposure by removing soiled articles of clothing and washing all exposed body parts, including the mouth and hair, as soon as possible.

Distance Rule 1: Do not remain near the site of the attack. Quickly put distance between yourself and potentially hazardous substances including smoke and debris.

Distance Rule 2: Direct contact is not required to receive a radiation exposure. Increased distance equates to decreased exposure to radiation.

Shielding Rule 1: In a radiation or biological attack, move to the dark corners of a basement (if available) or to a windowless center room. For attacks involving chemical agents, seek shelter in a windowless center room on the ground floor. The basement is where chemical agents will concentrate.

Shielding Rule 2: Heat the air of a sealed building to create positive pressure and prevent the infiltration of contaminants. Always use recirculated air or air purified by a HEPA filter. If this is not possible, turn the heating and air conditioning system off.

Shielding Rule 3: Use available resources to shield your lungs against airborne contaminants (e.g., cover your mouth with a handkerchief) and shield your body from radiation (e.g., move behind a concrete wall).

8.2 MINIMIZING EXPOSURE TO RADIATION FROM NUCLEAR EXPLOSION

The following survival rules describe what an individual can do to limit exposures to the intense heat from the fireball, initial radiation, and fallout following a nuclear explosion:

Nuclear Survival Rule 1: Close your eyes and turn away from the initial explosion. The intense heat from the fireball can cause temporary or permanent blindness.

Nuclear Survival Rule 2: Immediately seek shelter or retreat behind a solid barrier and stay hidden for at least 2 minutes. Every fraction of a second out in the open increases exposure to the intense heat, initial radiation, and air blast.

Nuclear Survival Rule 3: Get below ground if possible or move to the center of a large building before fallout arrives. The ground and layers of solid barriers (brick, cinderblock, concrete, etc.) will shield against radiation from fallout.

Nuclear Survival Rule 4: Stay below ground or in the center of a large building for up to 14 days, and do not leave unless reliable radiation measurements and instructions for evacuation are provided.

8.3 PREPARING FOR A NUCLEAR, CHEMICAL, OR BIOLOGICAL ATTACK

This section presents a number of preparatory measures recommended to help minimize exposure during a nuclear, chemical, or biological attack.

8.3.1 Emergency Preparedness

All schools, large urban buildings, grocery and department stores, shopping malls, entertainment centers, hospitals, government buildings, airports, subway and train stations, and sports stadiums are strongly encouraged to develop or to amplify existing emergency preparedness plans to address the appropriate responses to terrorist attacks involving weapons of mass destruction. Such plans should identify the specific measures that must be taken before, during, and after this type of incident. Having an emergency preparedness plan in place prior to an emergency and providing training to staff on the plan will ensure that everyone knows his specific roles and responsibilities.

All staff should receive initial training and annual refresher training on the contents of the emergency preparedness plan. Bringing in an emergency response specialist to assist with preparation of the emergency preparedness plan and provide staff training is an option that should seriously be considered.

Practice drills are an integral part of emergency preparedness training to ensure proper execution of the plan. These drills should be performed on an annual basis. Drills help refresh memories and can identify weaknesses in an emergency preparedness plan. Treat drills as seriously as an actual event to provide the best assurance of proper execution when the stakes are high.

8.3.2 Alarm Systems

All buildings (including residences) and sports stadiums require fire alarms or smoke detection systems. Schools, large urban buildings, hospitals, and government buildings should consider installing secondary alarm systems or intercom systems to alert staff and other occupants to hazards associated with weapons of mass destruction. When two alarm systems are installed in a building, one should be used to announce the evacuation; the secondary alarm should be used to direct occupants to remain inside the building until further notice. It is critical that the two alarm systems be easily distinguishable from one another, because responding incorrectly can have life-threatening consequences.

A second alarm system in a church, small business, grocery store, department store, shopping mall, entertainment center, airport, or subway or train station is *not* recommended because the public will automatically assume that any alarm sound is from a fire alarm and will exit the building. For these types of buildings, an intercom system should be considered to alert the public regarding the hazards present and how to respond. Using an intercom system instead of a second alarm system is also an acceptable option for schools, hospitals, and government buildings.

8.3.3 Air Purification Systems

Buildings should be equipped with means of ensuring the occupants have clean air to breathe. This can be accomplished by:

- Switching the heating/air conditioning system to recirculate mode to prevent contaminated outside air from entering the building
- Turning the heating and air conditioning system off
- Using a portable air purification system (e.g., the CARE 2000 air defense system) fitted with appropriate filters to remove contaminants from air already present in the building
- Using a permanent HEPA air filtration unit (e.g., the AirShelter bag-in/bag-out multistage filtration system) fitted with the appropriate filters to remove contaminants from the outside air before it is pulled into the building

8.3.4 Water Purification Systems

While an individual can survive on 2 to 3 cups of water per day, it is best to assume for planning purposes that an individual needs 1 gallon of water per day. A family or business should have a minimum of several days' water supply available for each member of the family or employee. However, a 2-week supply (14 gallons per person) is optimal. The simplest and likely the least expensive way to maintain a fresh water supply is to purchase bottled water at a local grocery store. The other option is purchasing a water filtration unit that can remove particulates or a unit that can remove everything ranging from radioactive and other particulates (e.g., anthrax spores), to metals, organics, herbicides, pesticides, bacteria, and parasites (e.g., the TGI Model 625U/DX).

8.3.5 Personal Protective Equipment

Personal protective equipment includes clothing and/or respiratory equipment that can be worn to protect the body against various forms of contamination. Some of the most common forms of personal protective equipment include dust masks, air-purifying respirators, protective suits made from particulates or chemically resistant materials (e.g., Tyvek), and lightweight protective rubber or chemical-resistant gloves.

Since dust masks are relatively inexpensive, it is recommended that all residences, businesses, and schools consider purchasing one or more dust masks for each family member, employee, and/or student. While an air-purifying respirator is much more expensive than a dust mask, with the appropriate filters it can provide protection from a multitude of airborne contaminants including radioactive dust particles, volatile radionuclides, anthrax spores, and chemical warfare agents. The advantage of a full-face respirator over a half-face respirator is that the full-face respirator also provides eye protection and generally provides a better seal against the face. These types of respirators should be fitted to the wearer's face and the appropriate type cartridge selected for the agents of concern.

Other relatively inexpensive personal protective items that are not absolutely essential but should be considered are protective garments and gloves. One-piece coveralls with head covers and booties made from lightweight plastic such as Tyvek are relatively inexpensive, semi-repellent, and disposable. Nitrile gloves are preferred over latex gloves because they have much greater tear and puncture resistance.

8.3.6 First Aid Kits

Well-stocked first aid kits are strongly recommended for all facilities and should include:

- Bandages (standard and butterfly types)
- Gauze pads (in multiple sizes)
- Gauze wraps
- First aid tape
- Sanitizing wipes

- Scissors
- Elastic wraps
- Arm sling
- Finger splints
- Eye pads
- Instant ice packs
- Antibacterial ointment
- Burn gel
- Eye flush
- Latex examination gloves
- Ibuprofen
- Vitamin supplements
- Other daily medications (e.g., pain relievers)
- Potassium iodine tablets (optional)

Potassium iodine tablets can be used to reduce radioactive iodine exposure to the thyroid gland. According to the National Council of Radiation Protection and Measurement (NCRP), taking 130 milligrams of potassium iodine at or before exposure to radioactive iodine effectively blocks nearly 100% of radioactive iodine from reaching the thyroid (1977). See Table 3-3 for a summary of antidotes for various chemical and biological agents.

8.3.7 Communication Devices

Because electrical lines may be knocked out by emergency events, it is recommended that one or more cellular telephones be maintained in each home, school, or business to be able to call for help and communicate with family members. Communicating with rescue workers and family during an emergency event will help minimize some of the psychological effects. Each home, school, or business should also maintain one or more battery-powered AM/FM or short-wave radios and several sets of replacement batteries. The radios can provide updated reports on emergency response efforts and other useful information.

8.3.8 Emergency Lighting

It is recommended that multiple battery-powered light sources (and replacement batteries and bulbs) or other forms of emergency lighting be maintained in all schools, large urban buildings, grocery and department stores, shopping malls, entertainment centers, hospitals, government buildings, airports, subway and train stations, and sports stadiums in the event that electrical power is out for an extended time. Backup lighting can help emergency response workers see where they are going, allow hospital staff to continue tending to critically ill patients, and prevent panic. Emergency lighting is optional for residences, churches, and small businesses. Emergency lighting consisting of one or more battery-powered camping-type lanterns or flashlights is usually sufficient for residences and small businesses.

8.3.9 Emergency Food Supplies

For residences, schools, and hospitals, it is recommended that an emergency food supply be available to feed family members, students, and patients for a minimum of several days and preferably for 2 weeks. Food should include nonperishable canned items such as tuna, soups, stews, vegetables, fruit, and fruit juices. The facilities should be supplied with disposable plates, bowls, cups, eating utensils (primarily spoons), multiple hand-operated can openers, and a large supply of plastic garbage bags and paper towels. Stocking an emergency food supply is optional for the other facilities.

8.3.10 Screening Instruments

It is strongly recommended that hospitals consider purchasing one or more hand-held radiation survey instruments (e.g., Ludlum Model 3 survey meter combined with Ludlum Model 44–9 pancake Geiger-Mueller detector) for identifying radiological contamination. In the event of a nuclear incident, one or more radiation survey instruments in a hospital emergency room would allow doctors to determine whether arriving patients are radiologically contaminated. They could provide doctors with information regarding the levels of radiation received and help control further spread of the contamination. It is also recommended that hospitals consider purchasing one or more chemical agent survey instruments (e.g., SAW Mini-CAD) or other type of agent detection method (e.g., colorimetric tubes, M18 and M256A1 detection kits, M8 and M9 detection paper).

8.4 GUIDANCE FOR EMERGENCY RESPONDERS

When an explosive device is used to disperse radioactive, chemical, or biological materials, the treatment of casualties is more difficult because of the presence of contamination. In this situation, emergency responders could face a life-threatening situation unless appropriate precautions are taken. These precautions include using screening instruments (see Section 6.12) to assess hazard conditions before responding, then selecting the appropriate level of personal protective equipment (see Section 6.7) to provide protection from the hazard.

Treatment of life-threatening injuries should almost always take precedence over measures to address radioactive, chemical, or biological contamination or exposure. The injured individuals should be stabilized if possible and immediately transported to a medical facility. It is recommended that an individual with training in the areas of radioactive, chemical, and biological hazards accompany the first patients to the hospital and serve as an advisor to the medical team.

The possibility of contamination may be determined in the field, en route to a treatment facility, or at the treatment facility, depending on the condition of the patient. Individuals subjected only to external contamination and not otherwise injured should be decontaminated (see Section 7.3) at a location other than a hospital. Patients who show no evidence of external contamination but have likely received internal contamination as a result of a wound or inhalation or ingestion of a contaminant do not need to be decontaminated before treatment. However, blood, vomit, urine, and feces from these patients may be contaminated and should be controlled as such.

Contamination assessment of an injured individual should be performed by a trained health professional under the supervision of on-the-scene medical personnel. This assessment should include collecting: radiation and chemical agent measurements using instrumentation identified in Section 6.12, nasal wipes, and collection of saliva, blood, and other samples for laboratory testing. The distribution of contamination and locations of wounds on the body should be recorded.

The use of volunteers for exposures during emergency actions is desirable. Older workers with low lifetime accumulated effective doses should be chosen from among the volunteers whenever possible. Where this cannot be accomplished, it is recommended that (from a radiological perspective) a limit of 50 rem effective dose and an equivalent dose of 500 rem to the skin be applied. When the equivalent dose to an emergency responder approaches or exceeds 50 rem, workers must understand the potential for acute effects and the increase in the lifetime risk of cancer.

Workers should also be selected based on their experience in performing required emergency tasks, since the time to accomplish tasks will likely be reduced and will help to minimize worker exposure. Only nonpregnant workers over the age of 18 should be selected.

Training of the emergency responders for handling situations involving weapons of mass destruction should emphasize critical concepts for self-preservation and effective casualty management. Training should include classroom instruction on basic concepts and principles, hands-on demonstrations of required skills, and drills to reinforce basic procedures.

Bibliography

40 CFR 61, National emission standards for hazardous air pollutants, *U.S. Code of Federal Regulations*, 1997.

42 USC 10101 et seq., Nuclear Waste Policy Act of 1982, Public Law 97-425, 1983.

42 USC 2011 et seq., Atomic Energy Act of 1946, Chapter 724, 60 Stat. 755, 1946.

42 USC 300f et seq., Safe Drinking Water Act of 1974, Public Law 93-523, 1974.

42 USC 7401 et seq., Clean Air Act, Chapter 360, 69 Stat. 322, 1955.

AEC, Radiation safety in atomic energy activities: staff report to the Atomic Energy Commission, *Twenty-First Semiannual Report*, U.S. Atomic Energy Commission, Washington, D.C.

AFRRI, *Medical Management of Radiological Casualties Handbook,* Special Publication 99-2, Armed Forces Radiobiology Research Laboratory, Bethesda, MD, 1999.

Ager, G.O. and Knights, L.M., *Fire Fighting Criteria for 234-5*, HW-76269, General Electric Company, Richland, WA, 1963.

Albright, E., Berhout, F., and Walker, W., *World Inventory of Plutonium and Highly Enriched Uranium*, Oxford University Press, New York, 1993.

ANS, *Controlled Nuclear Chain Reaction: The First 50 Years*, American Nuclear Society, LaGrange Park, IL, 1992.

Bair, W.J., Inhaled radioactive particles and gases, *Health Physics*, 10, 861, 1965.

Bair, W.J., *25th Annual Life Sciences Symposium: Radiation Protection, A Look to the Future*, Pergamon Press, New York, 1988.

Breitenstein, B.D., Medical management and chelation therapy, *Health Physics,* 45, 855, 1983.

Brodsky, A., Radioactive hazards in survival planning, *Radiation Protection Management*, 18, 34, 2001.

Brown, D., Weiss, J.G., Macvittie, T.J., and Pillai, M.V., *Treatment of Radiation Injuries*, Plenum Press, New York, 1989.

Burley, H.H., *Fuel Element Technical Manual*, HW-4000, General Electric Company, Richland, WA, 1956.

Bush, S.H., *Fuel Element Technical Manual*, HW-4000, General Electric Company, Richland, WA, 1956.

Bustad, L.K., Biology of radioiodine, *Proceedings of the Hanford Symposium on the Biology of Radioiodine*, Pergamon Press, Oxford, 1964.

Cantril, S.T. and Parker, H.M., Status of health and protection at the Hanford Engineer Works, in *Industrial Medicine on the Plutonium Project*, Stone, R.S., Ed., McGraw-Hill, New York, 1951.

Carlisle, R.P. and Zenzen, J., *Supplying the Nuclear Arsenal: American Production Reactors, 1942–1992*, Johns Hopkins University Press, Baltimore, 1996.

Caufield, C., *Multiple Exposures: Chronicles of the Radiation Age*, University of Chicago, Chicago, IL, 1989.

Collins, D.L., Stress at Three Mile Island: altered perceptions, behaviors and neuroendocrine values, in *The Medical Basis for Radiation-Accident Preparedness III: Psychological Perspectives*, Ricks, R.C. and Berger, M.E., Eds., Elsevier Science, New York, 1991.

Compton, J.A.F., *Military Chemical and Biological Agents*, Telford Press, Caldwell, NJ, 1987.

Conklin, J.J. and Walker, R.I., Eds., *Military Radiobiology*, Academic Press, New York, 1987.

Del Tredici, R., *At Work in the Fields of the Bomb*, Harper & Row, New York, 1987.

DeNeal, D.L., *Historical Events: Single Pass Reactors and Fuels Fabrication*, DUN-6888, Douglas United Nuclear Inc., Richland, WA, 1970.

DOE, *Human Radiation Experiments Associated with the U.S. Department of Energy and Its Predecessors*, DOE EH-0491, U.S. Department of Energy, Washington, D.C., 1995.

DOE, *Ordnance and Explosive Waste Records Search Report*, DOE/RL-94-07, Rev. 0, U.S. Department of Energy, Richland, WA, 1995.

DOE, *Plutonium: The First 50 Years, United States Plutonium Production, Acquisition, and Utilization from 1944 through 1994*, DOE/DP-0137, U.S. Department of Energy, Washington, D.C., 1996.

EPA, *Manual of Protective Action Guides and Protective Actions for Nuclear Incidents*, EPA 400-R-92-001, U.S. Environmental Protection Agency, Washington, D.C., 1992.

EPA, *Radiation Exposure and Risks Assessment Manual (RERAM): Risk Assessment Using Radionuclide Slope Factors*, EPA 402-R-96-016, U.S. Environmental Protection Agency, Washington, D.C., 1996.

Facer, J.F. and Weidenbaum, B., *The Recovery of Plutonium from Metal Wastes*, HW-22136, General Electric Company, Richland, WA, 1951.

Fermi, R. and Samra, E., *Picturing the Bomb: Photographs from the Secret World of the Manhattan Project*, Harry N. Abrams, Inc., New York, 1995.

Forbes, B.A., Sahm, D.F., and Weissfeld, A.J., *Bailey and Scott's Diagnostic Microbiology*, 10th ed., Mosby, St. Louis, 1998.

Ford, D., *The Cult of the Atom: The Secret Papers of the Atomic Energy Commission*, Simon & Schuster, New York, 1982.

Franc, D.R., *Defense against Toxin Weapons*, U.S. Army Medical Research Command, USAMRIID, Fort Detrick, MD, 1997.

Frigerio, N.A., *Your Body and Radiation*, Understanding the Atom Series, U.S. Atomic Energy Commission, Oak Ridge, TN, 1968.

Gaines, M., *Atomic Energy*, Grosset & Dunlap, New York, 1979.

GE, *Hanford Biology Research Annual Report for 1963*, HW-80500, General Electric Company, Richland, WA, 1964.

Gephart, R.E., *A Short History of Plutonium Production and Nuclear Waste Generation, Storage, and Release at the Hanford Site*, PNNL-SA-32153, Pacific Northwest National Laboratory, Richland, WA, 1999.

Gerber, M.S., *The Plutonium Production Story at the Hanford Site: Processes and Facility History*, WHC-MR-0521, Rev. 0, Westinghouse Hanford Company, Richland, WA, 1996.

Gilbert, E.S., Studies of workers exposed to low doses of external radiation, *Occupational Medicine*, 6, 665, 1991.

Gilbert, E.S. and Marks, S., Analysis of the mortality of workers in a nuclear facility, *Radiation Research*, 79, 122, 1979.

Gilbert, E.S. and Sever, L.E., *Draft Presentation Studies of Congenital Malformations and the Hanford Site*, PNL-10469-225, Pacific Northwest Laboratory, Richland, WA, 1987.

Glasstone, S. and Dolan, P.J., Eds., *The Effects of Nuclear Weapons*, 3rd ed., U.S. Department of Defense and U.S. Department of Energy, Washington, D.C., 1977.

Goans, R.E., Holloway, E.C., Berger, M.E., and Ricks, R.C., Early dose assessment following severe radiation accidents, *Health Physics*, 72, 513, 1997.

Gosling, F.G., *The Manhattan Project: Making the Atomic Bomb*, DOE/MA-0001-01/99, U.S. Department of Energy, Washington, D.C., 1999.

Groves, L.R., *Now It Can Be Told: The Story of the Manhattan Project*, Da Capo Press, New York, 1983.

HPS, *Radiation Risk in Perspective*, Position Statement of the Health Physics Society, adopted January 1996.

Hales, P.B., *Atomic Spaces: Living on the Manhattan Project*, University of Illinois Press, Chicago, 1997.

Hampton, J.D., The cell cycle in malignancy and immunity, *Proceedings of the 13th Annual Hanford Biology Symposium*, Atomic Energy Commission and Pacific Northwest Laboratory, Oak Ridge, TN, 1973.

Heid, K.R., *Emergency Radiological Plans and Procedures*, HW 70935, U.S. Atomic Energy Commission, Richland, WA, 1962.

Helgeson, G.L., *Further Notes on the Surface Dosage of Plutonium Metal*, HW-41369, General Electric Company, Richland, WA, 1956.

Hewlett, R.G. and Holl, J.M., *Atoms for Peace and War, 1953–1961: Eisenhower and the Atomic Energy Commission*, University of California Press, Berkeley, 1989.

Hill, O.F., *Symposium on Iodine Problem*, HW-39073, General Electric Company, Richland, WA, 1955.

Honstead, J.F., *Program for Evaluating Environmental Radiation Dose to Children*, BNWL-SA-1288, Pacific Northwest Laboratories, Richland, WA, 1967.

Honstead, J.F., Radionuclide burden–diet relationships near a nuclear facility, in *Diagnosis and Treatment of Deposited Radionuclides*, Kornberg, H.A. and Norwood, W.D., Eds., Excerpta Medica, Amsterdam, 1968.

Honstead, J.F. and Eichner, R.N., *Dietary and Body Burden Data and Dose Estimates for Local School Children and Teenagers*, Y80054, Pacific Northwest Laboratory, Richland, WA, 1972.

Hungate, F.P., Radiation and terrestrial ecosystems, *Health Physics*, 11, 1255, 1965.

IAEA, *Emergency Response Planning and Preparedness for Transport Accidents Involving Radioactive Materials*, Safety Series 87, International Atomic Energy Agency, Vienna, 1988.

IAEA, *Principles and Techniques for Post-Accident Assessment and Recovery in a Contaminated Environment of a Nuclear Facility*, Safety Series 97, International Atomic Energy Agency, Vienna, 1989.

IAEA, *Recommendations, Emergency Planning and Preparedness for Accidents Involving Radioactive Materials Used in Medicine, Industry, Research and Teaching*, Safety Series 91, International Atomic Energy Agency, Vienna, 1989.

IAEA, *Intervention Criteria in a Nuclear or Radiation Emergency*, Safety Series 109, International Atomic Energy Agency, Vienna, 1994.

IAEA, *International Basic Safety Standards for Protection Against Ionizing Radiation and for the Safety of Radiation Sources*, Safety Series 115, International Atomic Energy Agency, Vienna, 1996.

ICRP, *1990 Recommendations of the International Commission of Radiological Protection*, Publication 60, Pergamon Press, New York, 1991.

ICRP, *Pregnancy and Medical Radiation*, Publication 84, International Commission on Radiological Protection, Stockholm, 2000.

ICRP, *Principles for Intervention for Protection of the Public in a Radiological Emergency*, Publication 63, International Commission on Radiological Protection, Stockholm, 1993.

ICRP, *Protection of the Public in the Event of Major Radiation Accidents: Principles for Planning*, Publication 40, International Commission on Radiological Protection, Stockholm, 1984.

ICRP, *Recommendations of the International Commission on Radiological Protection*, Publication 26, International Commission on Radiological Protection, Stockholm, 1977.

ICRP, *Recommendations of the International Commission of Radiological Protection*, Publication 60, International Commission on Radiological Protection, Stockholm, 1991, 1990.

ICRP, *1990 Recommendations of the International Commission of Radiological Protection*, Publication 60, International Commission on Radiological Protection, Stockholm, 1991.

IOM/NRC, *Chemical and Biological Terrorism: Research and Development to Improve Civilian Medical Response*, Institute of Medicine/National Research Council, National Academy Press, Washington, D.C., 1999.

Kathren, R.L., The United States transuranium and uranium registries: 1968–1993, *Radiation Protection Dosimetry*, 60, 349, 1995.

Kathren, R.L. and Ziemer, P.L., Eds., *Health Physics: A Backward Glance*, Pergamon Press, New York, 1980.

Lambert, B.E., The adequacy of current occupational standards for protecting the health of nuclear workers, *Occupational Medicine*, 6, 725–740, 1991.

Loeb, P., *Nuclear Culture: Living and Working in the World's Largest Atomic Complex*. Coward, McCann & Geoghegen, New York, 1982.

Macvittie, T.J., Weiss, J.G., and Browne, D., Advances in the treatment of radiation injuries, in *Advances in the Biosciences*, Vol. 94, Elsevier Science, New York.

Mancuso, T.F., Stewart, A., and Kneale, G., Radiation exposures of Hanford workers dying from cancer and other causes, *Health Physics*, 33, 369–385, 1977.

McMaster, B.N. et al., *Historical Overview of the Nike Missile System*, DRXTH-AS-IA-83016, U.S. Army Toxic & Hazardous Materials Agency, Aberdeen Proving Grounds, MD, 1984.

Mettler, F.A., Jr. and Upton, A.C., *Medical Effects of Ionizing Radiation*, 2nd ed., W.B. Saunders, Philadelphia, 1995.

Murray, P.R., Baron, E.J., Pfalles, M.A., Tenover, F.C., and Youken, R.H., *Manual of Clinical Microbiology*, 6th ed., ASM Press, Washington, D.C., 1995.

NCRP, *Advising the Public about Radiation Emergencies: A Document for Public Comment*, Commentary 10, National Council on Radiation Protection and Measurements, Bethesda, MD, 1994.

NCRP, *Limitation of Exposure to Ionizing Radiation*, Report 116, National Council on Radiation Protection and Measurements, Bethesda, MD, 1993.

NCRP, *Management of Persons Accidentally Contaminated with Radionuclides*, Report 65, National Council on Radiation Protection and Measurements, Bethesda, MD, 1980.

NCRP, *Management of Terrorist Events Involving Radioactive Materials*, Report 138, National Council on Radiation Protection and Measurements, Bethesda, MD, 2001.

NCRP, *Protection of the Thyroid Gland in the Event of Releases of Radioiodine*, Report 55, National Council on Radiation Protection and Measurements, Bethesda, MD, 1977.

NCRP, *Radiological Factors Affecting Decision-Making in a Nuclear Attack*, Report 42, National Council on Radiation Protection and Measurements, Bethesda, MD, 1974.

NCRP, *Recommended Screening Limits for Contaminated Surface Soil and Review of Factors Relevant to Site-Specific Studies*, Report 129, National Council on Radiation Protection and Measurements, Bethesda, MD, 1999.

NCRP, *Risk Estimates for Radiation Protection*, Report 115, National Council on Radiation Protection and Measurements, Bethesda, MD, 1993.

NCRP, *The Control of Exposure of the Public to Ionizing Radiation in the Event of Accident or Attack*, Symposium Proceedings 1, National Council on Radiation Protection and Measurements, Bethesda, MD, 1982.

NCRP, *Uncertainties in Fatal Cancer Risk Estimates Used in Radiation Protection*, Report 126, National Council on Radiation Protection and Measurements, Bethesda, MD, 1997.

Newton, C.E., Jr. et al., *Tissue Sampling for Plutonium through an Autopsy Program*, BNWL-SA-918, Battelle Memorial Institute and Pacific Northwest Laboratories, Richland, WA, 1966.

Northrup, J., *Handbook of Nuclear Weapon Effects*, Defense Special, Weapons Agency, Alexandria, VA, 1996.

Norwood, W.D., Planning for care of injured radiation patients, *Archives of Environmental Health*, 8, 474, 1964.

Phillips, R.D., Biological effects of extremely low frequency electromagnetic fields, *Proceedings of the 18th Annual Hanford Life Sciences Symposium*, U.S. Department of Energy, Pacific Northwest Laboratory, and Electric Power Research Institute, Oak Ridge, TN, 1978.

Pickett, M., *Explosives, Identification Guide*, Delmar Publishers, New York, 1999.

Rhodes, R., *Dark Sun: The Making of the Hydrogen Bomb*, Simon & Schuster, New York, 1995.

Shleien, B., *The Health Physics and Radiological Health Handbook*, rev. ed., Scinta, Silver Springs, MD, 1992.

Sidell, F.R., Patrick, W.C., and Dashiell, T.R., *New York State Chem-Bio Handbook*, James Information Group, Alexandria, VA, 2000.

Sloat, R.J., *Critical Mass Control Specification: Transporting of Plutonium-Bearing Materials by Motor Truck*, HW-74716, General Electric Company, Richland, WA, 1963.

Smyth, H.D., *Atomic Energy for Military Purposes*, Stanford University Press, Stanford, CA, 1989.

Stannard, J.N., *Radioactivity and Health: A History*, DOE/RL/01830-T59, Pacific Northwest Laboratory, Richland, WA, 1988.

Thompson, R.C., The biology of the transuranic elements, *Health Physics*, 8, 561, 1962.

Thompson, R.C. and Blair, W.J., The biological implications of the transuranium elements, *Proceedings of the 11th Hanford Biology Symposium*, Pergamon Press, London, 1972.

Thompson, R.C. and Mahaffey, J.A., Life-span radiation effects studies in animals: what they can tell us, *Proceedings of the 22nd Hanford Life Sciences Symposium*, U.S. Department of Energy and Pacific Northwest Laboratory, Oak Ridge, TN, 1983.

Tierno, P.M., Jr., *Protect Yourself against Bioterrorism: Everything You Need to Know about Anthrax, Plague, Botulism, Smallpox, Encephalitis, Cholera, Hemorrhagic Fevers, Ricin, and More*, Pocket Books, New York, 2002.

Tierno, P.M., Jr., *The Secret Life of Germs*, Pocket Books, New York, 2001.

USAMRIID, *Medical Management of Biological Casualties*, 3rd ed., Fort Detrick, MD,1998.

Wilkinson, G.S., Epidemiologic studies of nuclear and radiation workers: an overview of what is known about health risks posed by the nuclear industry, *Occupational Medicine*, 6, 715, 1991.

[illegible] Reston, Virginia, VA 19[illegible].

[illegible] planning [illegible] 1965.

[illegible]

[illegible] Information Group [illegible]

[illegible] Publishers [illegible]

[illegible] Richland, WA, 19[illegible].

[illegible] The history of the [illegible]

[illegible]

[illegible] Health Physics [illegible]

Waltar, A.E. America the Powerless [illegible]

[illegible]

Wilson, R. [illegible]

Index

C

D

E

F